Amazonite: Mineralogy, Crystal Chemistry, and Typomorphism

Amazonite: Mineralogy, Crystal Chemistry, and Typomorphism

Mikhail Ostrooumov

Institute of Earth Sciences, University of Michoacan of
San Nicolas of Hidalgo, Morelia, Mexico

ELSEVIER

AMSTERDAM • BOSTON • HEIDELBERG • LONDON • NEW YORK • OXFORD
PARIS • SAN DIEGO • SAN FRANCISCO • SINGAPORE • SYDNEY • TOKYO

Elsevier
Radarweg 29, PO Box 211, 1000 AE Amsterdam, Netherlands
The Boulevard, Langford Lane, Kidlington, Oxford OX5 1GB, UK
225 Wyman Street, Waltham, MA 02451, USA

This English language edition is a translation of original Russian language edition titled
Амазонский камень, ISBN 978-5-7325-0675-4
Copyright notice: © Издательство "Политехника", 2008.

Notices
Knowledge and best practice in this field are constantly changing. As new research and
experience broaden our understanding, changes in research methods, professional
practices, or medical treatment may become necessary.

Practitioners and researchers must always rely on their own experience and knowledge
in evaluating and using any information, methods, compounds, or experiments
described herein. In using such information or methods they should be mindful
of their own safety and the safety of others, including parties for whom they have a
professional responsibility.

To the fullest extent of the law, neither the Publisher nor the authors, contributors, or
editors, assume any liability for any injury and/or damage to persons or property as a
matter of products liability, negligence or otherwise, or from any use or operation of
any methods, products, instructions, or ideas contained in the material herein.

ISBN: 978-0-12-803721-8

British Library Cataloguing in Publication Data
A catalogue record for this book is available from the British Library

Library of Congress Cataloging-in-Publication Data
A catalog record for this book is available from the Library of Congress

For Information on all Elsevier publications
visit our website at http://store.elsevier.com/

Working together
to grow libraries in
developing countries

www.elsevier.com • www.bookaid.org

Cover photo: Amazonite open pit on the Ploskaya Mountain, Western Keivy, Kola
Peninsula, Russia. Photo: Natalia A. Pekova, 1996. From archives by Igor V. Pekov.

Contents

Preface

This book is the synthesis of amazonite mineralogical investigation and the latest results of modern studies of this famous mineral in their most sophisticated aspects.

The first edition of this monograph (*Amazonit,* [Nedra Publishing House, 1989]) became a bibliographic rarity a few years after its publication. Therefore the author, M.N. Ostrooumov, has prepared a second edition of the volume, which takes into account the latest data and introduces a number of necessary additions and corrections to the previous version.

Amazonite has been the subject of a significant quantity of research that treats its various attributes: crystal chemistry, properties, conditions of formation, genesis, prospecting significance, and others. However, there is no exhaustive publication that offers a comprehensive review of the aforementioned questions while providing an insightful analysis of the often contradictory research opinions on amazonite.

The goal of this work is to provide a full consideration of amazonite that includes information useful in theory and in practice to specialists from a wide variety of fields. In particular, at present it is very important to discuss the practical problem about the use of amazonite in prospecting for deposits of rare metals and rare-earth elements (REE).

The present research is based on the materials collected by Ostrooumov et al. [47] at multiple amazonite deposits in Russia and other countries, the analysis of a multitude of published sources on all known aspects and occurrences of this mineral, and the results of experimental studies.

The fieldwork conducted by Ostrooumov et al. [47] has covered virtually every significant amazonite deposit and occurence in the Kola Peninsula, the Urals, Kazakhstan, Eastern Siberia, Transbaikal, Karelia, Central Asia, and Ukraine, as well as a range of deposits in Brazil, India, Canada, Mexico, Mongolia, USA, and other countries.

The goal of the conducted laboratory research was to obtain the fullest possible description of crystal chemical features and properties of amazonite. In addition to the usual mineralogical and petrographic research methods (studies of hand specimens and thin sections), Ostrooumov et al. [47] employed a variety of special methods including optical spectrometry in wide spectral range, colorimetry, vibrational spectroscopy, electron paramagnetic resonance (EPR), X-ray diffraction, X-ray, photo- and thermoluminescence, heat treatments, and X-ray irradiation experiments.

Full wet chemical analysis, partial X-ray fluorescence analysis, and optical spectrographic analysis of amazonite and associated potassium feldspars were performed in the following Russian and international laboratories: A.P. Karpinsky Russian Geological Research Institute (VSEGEI), PGO Sevzapgeologia, MNTK Mekhanobr, Ilmen State Nature Reserve, Ukraine Academy of Sciences Institute of Geochemistry and Mineral Physics, Mineralogical Institute (Mainz University, Germany), Materials Institute (University of Nantes, France), Geological Institute (Mexico National University) and Institute of Earth Sciences (University of Michocan de San Nicolas de Hidalgo, Morelia, Mexico).

In preparing the revised second edition of this monograph, M.N. Ostrooumov conducted additional research of amazonite in the aforementioned laboratories of the German, French, and Mexican universities, including but not limited to Raman spectroscopy, EPR, and X-ray irradiation.

The comprehensive approach to the studies of amazonite represented in this monograph was made possible largely by the cooperative efforts of Russian and international specialists working in various areas of mineralogy, who had researched various aspects of this mineral in previous years.

This second edition has been prepared in its entirety by M.N. Ostrooumov, who also prepared for publication the first edition of this volume in 1989. As the author of the present edition (Preface, Introduction, Chapters 1–6, and Conclusion), M.N. Ostrooumov would like to express his gratitude to the following colleagues for their cooperation in the preparation of the second edition: the collaboration with A.N. Platonov (Sections 4.4.2 and 5.1) and V.A. Popov (Sections 2.4 and 4.1) has been most fruitful. The author is much obliged to all of the organizations that assisted in bringing the second edition of this work up to a new level on par with the latest achievements in mineralogy: the National Mineral Resources University in St Petersburg, Ilmen Nature Reserve (Russia), Mainz University (Germany), University of Nantes (France), University of Michoacan de San Nicolas de Hidalgo (Mexico), and others.

Introduction

Amazonite is a mineral that has attracted scientific attention for generations and has been studied by prominent geologists including A. des Cloiseaux, N.I. Koksharov, V.I. Vernadsky, A.E. Fersman, V.M. Goldschmidt, and A.N. Zavaritsky. The history of amazonite discovery and scientific research is rich in riddles, paradoxes, misconceptions, and presuppositions. The time and place of the first amazonite finding is heavily debated even today. The very name of the mineral seems paradoxical, as there have been no known amazonite fields in the Amazon River basin. Evaluations of the mineral's significance remain quite controversial [1, 2, 23]. Despite more than two centuries of research history and innumerable papers dedicated to amazonite, it continues to draw the attention of geologists and mineralogists. What are the reasons for this continuing interest in amazonite?

The most noticeable aspect for any geologist encountering amazonite in the field or in the lab is certainly the mineral's color: a rich palette of blue and green, characteristic of such rare and precious gems as turquoise, malachite, emerald, and aquamarine. However, this beautiful mineral with a color unique for a semiprecious stone has attracted the attention of a wide range of specialists mainly thanks to the discovery that it is not as rare as had been supposed previously. Amazonite is a variety of potassium feldspar, a group of minerals extremely widespread in the Earth's crust. But it is due to its color that amazonite stands out among other potassium feldspars.

Scientists have undertaken numerous attempts to understand and decipher the nature of this mineral's color. A wide variety of opinions have successively replaced one another as new characteristics of amazonite composition, crystal chemistry, and properties were discovered. In the last decades, the tendency has been to connect the chemical and structural characteristics of amazonite, and yet there is no hypothesis that would account for the diversity of its properties and crystal chemistry.

Until recently, amazonite was considered a mineralogical rarity not worthy of special attention and usable only as a semiprecious stone. The few amazonite fields known at that time were linked to certain types of Precambrian and Paleozoic granitic pegmatites. These pegmatites were noted by many researchers to have a particular composition of amazonitic paragenesis: amazonite associates with albite, topaz, beryl, fluorite, tourmaline, micas, and a wide range of rare-earth and rare-metal minerals. It is noteworthy that in the previous century, miners commonly considered amazonite a sure sign for finding topaz (known in Russian as a heavyweight): "Extensive experience taught miners to value this stone highly as the best sign for finding a 'heavyweight' (there are mines with amazonite but without topaz, but the reverse correlation had apparently never been observed). They know very well that the more intense the amazonite color, the greater the chance to hit a lucky vein" [24]. Thus A.E. Fersman formulated this idiosyncratic prospecting rule that could be considered one of the first mineralogical criteria for pegmatite prospecting and evaluation. The formulation of this rule resulted in the contradictory situation that remains unsolved up to the present day: some researchers are very optimistic regarding the significance of amazonite as a search indicator of a certain mineral complex, while others remain skeptical of this potential use of amazonite. The negative conclusions were based mainly on occurrences of pegmatites similar in type to those containing amazonite, which display rich mineralization but no amazonite, as well as occurrences of amazonitic pegmatites with subeconomic or accessory mineralization.

A.N. Zavaritsky's classic works on Ilmen Nature Reserve pegmatites contributed to the shifting of views on the narrowly mineralogical significance of amazonite color. The color was recognized by a range of geologists as an indicator of a certain metasomatic phenomenon causing a characteristic secondary color of microcline—the effect later named *amazonitization* after the suggestion of A.N. Zavaritsky.

The discovery of this phenomenon allowed a new, deeper approach to understanding the nature of amazonite color, including not only its cause (a presupposed presence of one or another isomorphic impurities in microcline), but also the time, modes, and conditions of its formation. Important observations made by A.N. Zavaritsky and other geologists connected amazonitic microcline color with albite and quartz development zones.

However, contradictory opinions still abound regarding the essence and meaning of the amazonitization process. Not all researchers accept even the mere possibility of secondary amazonitic color. The followers of A.E. Fersman, for instance, still consider crystallization of residual melt or fluid enriched with volatile components and rare elements as the only possible modes of amazonite formation. In relation to the potassium feldspars with amazonite color, the term *K-feldsparing* is used in addition to A.N. Zavaritsky's term amazonitization. These cases have never been considered closely to determine the modes by which the color was formed, which could be either primary or secondary.

In addition, attempts have been made to substantiate the idea of de-amazonitization, that is, amazonite decoloration, as an alternative to amazonitization (secondary formed color). This has provided yet another possible explanation for the knotty pattern in the color of amazonite from certain deposits, but the question remained whether amazonite color is primary or secondary.

Discussions of this genetic problem might have remained a side issue of pegmatite genesis were it not for the mid-twentieth-century discoveries of large plutons of amazonitic granites and entire provinces of amazonite-bearing rock. It is worth noting that some of those areas were found to contain a minable concentration of a number of rare-metal minerals. Moreover, the conducted evaluations enabled some varieties of amazonitic granites to be classified as useful only for facing stone, and others as potentially promising raw material for the assemblage of lithophilous rare elements. As in pegmatite deposits, granite amazonitization to scale turned out to be commensurate with and closely linked to the processes of albitization, greisenization and ore formation. As a direct result, discussion has flared anew about the practical use of amazonite as a prospect indicator of certain mineral deposits and about the comparability of the petrological significance of amazonitization with that of other post-igneous processes. It is worth recalling here that some time ago an analogous discussion was taking place about the characteristics of granitoid albitization, which, as was discovered subsequently, occurs during the late post-igneous stages of granite intrusion formation. It has a variety of structural and morphological manifestations and varied ore-controlling significance.

At present, both the geochemistry and the distinctive features of amazonitization remain only partially clear. Multifaceted research of amazonite-bearing paragenesis would allow the description of the chemistry of all minerals contained therein. According to a range of research data, in comparison to associated potassium feldspars of common (non-amazonitic) color, amazonite displays higher concentrations of rubidium, cesium, plumbum, thallium, and certain other elements. However, despite the large number of papers treating the subject of the chemical composition of amazonite, the precise limits remain unknown as to the isomorphic capacity of this mineral in relation to rare and trace elements and their typical concentrations in amazonites contained in various genetic types of rocks. The aforementioned characteristics of amazonitic feldspar allow it to be classified as highly promising for determining the absolute age of the amazonite-bearing rock by means of three basic types of isotope analysis: potassium-argon, rubidium-strontium, and uranium-plumbum. The consistency of results must be checked against a wide range of identifiable ages.

The effusive analogues of Mesozoic amazonitic granites (so-called ongonites) as well as Alpine amazonitic pegmatites discovered in recent decades enable us to speak of a wide range of geological environments where amazonite can be found and highlight the necessity of discussing the typomorphic significance of this mineral.

The polemical character of the views on amazonite is the result of the wide variety of fragmentary and contradictory facts scattered throughout the published works and left without explanation and systematization. The state of studying amazonite becomes clear when one considers that even the most widely studied features of the mineral—its geological setting and characteristic color, its spectroscopic and colorimetric evaluation—have not received enough attention in the literature. It is, therefore, not surprising that the necessary and sufficient conditions for amazonite color have not been formulated with any degree of certainty, and consequently, it has not been possible to accurately predict the possibility of amazonite formation in various rocks, fields, and regions. Not one of the numerous works dedicated to amazonite has concentrated on establishing a comprehensive investigation of this mineral. The research data that would concentrate on typomorphic characteristics of amazonite is absent in the literature. There is insufficient information on the geological setting, paragenesis, ontogeny, and the nature of amazonitic feldspar color, and the very concept of amazonite requires revision and elaboration.

The author of this monograph is convinced that the variety of contradicting opinions on various aspects of amazonite research is the result of a vague understanding of amazonite's geological setting. A detailed analysis of the latter is essential for providing accurate information for the theoretical and practical aspects of amazonite research. On this account, the present work is first of all a compilation and a summary of the previously known and newly received data on the geological setting, structure, mineralogical, geochemical, and other distinctive features of amazonite-bearing rocks. The resulting analysis along with the close inspection of the features of amazonite from a variety of genetic types of rocks and formations allows a new approach to the study of amazonite in both its theoretical and practical aspects.

At the present time, amazonite has been discovered on every continent except Antarctica. There are over a 100 large amazonite deposits where, as a rule, it is the main rock-forming mineral. Thus, it may be taken as an established fact that this mineral is not as rare as previously considered. The list of amazonite discovery regions (in Russia and in other countries) includes more than 200 places, some of which include entire provinces of amazonite-bearing rock (e.g., the Kola Peninsula, Il'menskie Mountains, and Transbaikal). For instance, one pegmatite field in the Kola Peninsula (and this area contains several such fields) encompasses more than a hundred pegmatite veins with amazonite. There are just as many amazonite-bearing pegmatites in Ilmen vein field, in the Southern Urals, and in the Northern Baikal regions. A few dozen fields of amazonitic granites have been discovered in Transbaikal, Kazakhstan, Mongolia, and other regions. These facts once again prove the necessity to reevaluate the outdated ideas on the distribution areas of amazonite and its significance in contemporary mineralogy.

In their field research, the author et al. [47] (M.N. Ostrooumov and V.A. Popov) paid special attention to studying the geological conditions of occurrence, structure, and mineral composition of amazonite-bearing rock, as well as the main paragenesis, ontogeny specifications, and localization conditions of various amazonite generations. The data provided by this research supports the conclusion that the processes of albitization, greisenization, and ore mineralization are genetically interconnected. This raised the importance of discussing the introduction of the term amazonitization, analyzing the thermodynamic, physical and chemical conditions of amazonite formation, and investigating the primary or secondary nature of amazonite color.

The main goals of amazonite laboratory research included determining the typomorphic features of the mineral found in various genetic types of rock, its isomorphic capacity in relation to rare elements, its structural state, the nature of its color, its internal structure, and a quantitative evaluation of its color.

The author et al. [47] conducted amazonite research with the goal of obtaining information most relevant to the current problems of contemporary mineralogy. The detailed research conducted on the crystal chemistry and properties of amazonite has resulted in well-grounded judgments about the genesis and age of amazonite-bearing rocks as well as in the ability to predict the best search of ore mineralization. This has once again proven the importance of mineralogical prospecting and evaluation methods. Considering the reevaluation of the natural occurrence of the mineral and the fact that it contains a range of rare isomorphic elements, amazonite has the potential to become a new source of some elements and a reliable geochronometer.

The understanding of the versatility of amazonite that forms the basis of this work has been extremely conducive to forming a new perspective on the connection between its properties, crystal chemical features, and geological setting. This also helped to provide an improved interpretation of the nature of amazonite color and to evaluate its potential use as a typomorphic, prospecting, and evaluation indicator.

Research History of Amazonite

1 —

The rich blue-green colors of amazonite have been the object of mineralogists' and geologists' investigations for over 200 years. Among the many hypotheses attempting to explain the cause of this specific color, none takes into account all of the crystal chemical features of amazonitic K-feldspar. It is clear that the color of amazonite is affected by a number of parameters, which reflect the great variation in the chemical and structural peculiarities of this variety of potassium feldspar. During the last few decades, geologists and mineralogists have discussed the use of amazonite in the exploration for deposits of rare metals and rare-earth elements, and there are differing positive and negative opinions regarding this problem.

1.1 DISCOVERY AND RESEARCH CONDUCTED BEFORE THE END OF THE NINETEENTH CENTURY

In the era of qualitative-descriptive mineralogy stretching from antiquity up to the end of the eighteenth century, knowledge of feldspars—the primary rock-forming minerals—was extremely scant. In mineralogical tracts from the end of this period, feldspars were differentiated only by color, using fragmentary, qualitative, and frequently imprecise data regarding their chemical composition.

Specimens of feldspars collected by naturalists from a range of deposits in Europe and Russia served as the material basis for the first quantitative observations and generalizations obtained through the methods of chemistry and crystallography. Among the specimens discovered and researched, almost all of these traveling geologists noted the green feldspar from Chebarkul' (Il'menskie Mountains).

Soon after the discovery of the chemical element potassium by M.H. Klaproth in 1797, G. Vokelen completed the chemical analysis of the green feldspar of Siberia and established its membership among the potassium varieties. In 1801,

R.J. Haüy presented the first formulation of the composition of potassium feldspars in his famous work *Mineralogy*. Already known at that time were the results of J.J. Bindheim's still earlier chemical analysis of green feldspar, the source location of which was not indicated; this feldspar contained a copper impurity that sufficed as a simple explanation for the mineral's color. Although copper was not identified in G. Vokelen's analysis, A. Breithaupt, and later in 1866, N.I. Koshkarov, citing the research of K.F. Plattner, likewise considered copper responsible for amazonite color. The authority of these researchers served, henceforth, as the reason for the many attempts to detect a copper impurity in amazonite. Looking ahead, we should note that it was only in 1969 that researchers discovered a perfectly distinct phase of blue-green feldspar, called plagioclase-amazonite, the color of which actually was determined to be caused by a copper impurity. It is possible that the specimens that had been analyzed in the eighteenth century represented similar plagioclases.

By the nineteenth century, after numerous failed attempts to detect copper in amazonite, certain researchers discarded the opinion that the presence of this element was the reason for the stone's color. In 1876, the prominent mineralogist A. Des Cloizeaux was the first to note amazonite's tendency to lose color under heating to the point of incandescence. "These circumstances, as well as the constant loss under direct heating observed in the analyses, serve as definite evidence that the color of amazonstone is imparted by certain organic substances"—thus Des Cloizeaux concluded from his findings in 1891. G.G. Lebedev also asserted in his *Mineralogy* that the green color of amazonite is not caused, as was previously thought, by an impurity of a small quantity of copper oxide. The presence of a minute quantity of organic matter in amazonite was confirmed by K.K. Matveev in 1947 in the course of specialized experiments. However, not one of these works (as well as later works, e.g., V.N. Frolovskii) contained direct evidence of the causation of amazonite color by bitumen impurity.

Thus, by the end of the nineteenth century, the first of the prominent hypotheses for the reason of amazonite color had already been discredited to a significant degree, while the organic hypothesis advanced in its place, likewise, remained to be authoritatively proven. Remaining unquestioned was only the classification of amazonite among the potassium alkali feldspars.

Dating to this period are the first investigations of another important particularity of the constitution of amazonite: its crystal structure, which was studied from a crystallographic perspective. Prior to the 1820s, all feldspathic species known at the time, including amazonite as well as labradorite, adular, and albite, were classified as monoclines. In 1801, R.J. Haüy proposed for them the general term orthoclase, derived from the mineral's tendency to fracture along a right angle. The first measurement of the corners between the cleavage planes of feldspars, performed in 1823 by G. Rose, enabled the discovery of the triclinic symmetry of albite, labradorite, and anorthite, and separated the latter from the feldspars proper, which, in particular, included amazonite. In 1817, F. Breithaupt divided all feldspars into orthoclasic and plagioclasic, assigning amazonstone which he named amazonite, to the latter group. In 1830, that same mineralogist described a green feldspar from Greenland that did not fracture at a right angle and was thus called a microcline.

In 1866, N.I. Koshkarov in his prominent work *Materials for the Mineralogy of Russia* defined amazonite as a phase of orthoclase and, furthermore, citing the research of A. Des Cloiseaux, wrote: "… not all crystals of amazonstone belong without exception to the monoclinic system; on the contrary, some of them belong to the triclinic system. All crystals having deep green color and opacity belong to the triclinic system; on the contrary, the crystals of amazonstone are rather transparent and green in parts—the essence of monoclinic crystals." These observations, remaining either unnoticed or underappreciated by researchers of that time, in essence anticipated the conclusions reached a century later regarding the connection of amazonite color to the stone's structural particularities.

Subsequently, in his research of a specimen of amazonite from Murzinka, A. Des Cloiseaux established precisely its triclinic optical orientation and (after A. Breithaupt) distinguished definitively the microcline as an independent type of potassium feldspar. Thus, in the last quarter of the nineteenth century, the presentation of amazonite as regards the triclinic modification of potassium feldspars was firmly established in the literature.

1.2   STUDIES OF AMAZONITE IN THE FIRST HALF OF THE TWENTIETH CENTURY

In 1913, V.I. Vernadsky was the first to bring attention to the high rubidium content (up to 3.12% Rb_2O) in Il'menskie amazonite, noting, however, that some orthoclases are still richer in this element.

The data obtained by V.I. Vernadsky found confirmation much later—in 1935, Iu.M. Tolmachev and A.N. Filipov, having studied the chemical composition of amazonites from various deposits in the Urals, Kola Peninsula, Madagascar, and Colorado Plateau, detected in them the presence of notable quantities of rubidium, which strongly varied according to the source deposit of the stone. Regarding this as a characteristic particularity of amazonite, the authors, moreover, observed traces of lead in all specimens, but they did not find any link between the lead and color.

On the basis of results obtained from his own research in 1938, V.M. Goldschmidt with collaborators classified the presence of rubidium in amazonite as a necessary condition for the phenomenon of the green color of potassium feldspar. Still later, in 1954, V.M. Goldschmidt, abandoning his previous hypothesis, proposed that amazonite color might be caused by thallium atoms or ions activated by natural radiation.

In 1939, N.P. Kapustin identified a correlation between the intensity of color of a given amazonite and its rubidium content; unfortunately, this discovery was accepted by certain researchers (K.K. Zhirov et al., G.S. Pliusnin) as little more than a curiosity. It should be noted that N.P. Kapustin measured the intensity of color in terms of its maximum reflection (or transparency). For that period, this methodology was fully acceptable, so this finding was considered unquestionable experimental evidence. Subsequently, Kasputin's work was cited more than once as the evidential basis for a correlation between the color of amazonite and its rubidium content. From the point of view of contemporary understanding of the nature of mineral color, the role of rubidium as a center or precursor site of color is unlikely; however, the presence of rubidium in amazonite is not accidental and, as will be shown below, does hold a place in the aggregate of reasons that determine the amazonitization phenomenon as a whole.

At this time, progress quickly developed on new aspects of the amazonite subject: geological, genetic, and applied.

The sources of new interest in amazonite can be dated to the nineteenth century. Precisely during this period of intensive development of the Il'menskie amazonite mines, a close connection was noted between amazonite and topaz, along with rare black minerals (columbite, fergusonite) as well as rare and specific minerals such as cryolite and chiolite. Among the miners in the Il'menskie region, the prospecting significance of amazonite was well known as a reliable indicator of semiprecious stones.

From the first decades of the twentieth century, scientific research began of the Il'menskie Mountains. In 1928, E.O. Kopteva-Dvornikova described the mineral composition and characteristics of the internal texture of topaz-amazonite veins, noting in particular that in the transition zone from graphic granite to the central, interfolded amazonite and quartz, in the majority of cases albite is widespread, and there are often concentrations of rare-earth metals and minerals with volatile components.

The first synthesis of the geology, mineralogy, and genesis of amazonite was produced by A.E. Fersman [23]. In the classification he developed for granite pegmatites, amazonites are registered among the first four types together with the minerals fluorine, beryllium, and boron, and rare elements such as uranium, niobium, tantalum, and yttrium. He established the definitive characteristics of the composition and geological position of these pegmatites (see Chapter 5.2). According to Fersman's hypothesis, amazonite in pegmatites crystallized from melting solution in one of the late (pegmatoid) geophases, that is, at temperatures of 600 to 500 °C. Furthermore, he indicated that not infrequently "the formation of amazonstone occurs in earlier or later phases—the graphic granite and pneumatylitic phases, for which formation temperatures are respectively 700–600 °C and 500–400 °C."

A new point of view regarding the genesis of amazonite, fundamentally distinct from that mentioned above, was proposed and developed by A.N. Zavaritsky [6]. In his description of the amazonite mines of the Ilmen reserve, he devoted special attention to the character of the distribution of the variously colored sections of microcline, emphasizing the occurrence of a gradual transition from the usual potassium feldspars to amazonite, while noting that the crystals of the latter possess a bright color only on a side facing a cavity or a vein of quartz. In his opinion, this results in the impression of a subsequent change in the color of a feldspar under the action of residual dissolution, feldspar's own version of the amazonitization process. Based on the experimental data obtained previously by N.P. Kapustin, A.N. Zavaritsky interpreted amazonitization as the process of ionic metasomatic substitution of part of a microcline's potassium ions by rubidium ions.

In 1940 on the southwest shores of Lake Balkhash, the geologists S.I. Letnikov and B.S. Dmitrievsky described a previously unknown genetic type of amazonite-containing rock: amazonitic granite. This discovery and especially the resulting series of subsequent findings stimulated a new rise of interest in amazonite among geologists. As follows from Letnikov and Dmitrievsky's description, amazonitic granites are associated with a series of their vein retinue: amazonitic granite-aplites, aplites, and pegmatites. In the opinion of these pioneers, amazonitic granites should interest geologists as much for their complex of rare-metal ores as for their possible use as a beautiful cut stone.

1.3 RESEARCH OF RECENT DECADES

In this section, we will first examine the results of various works that discuss amazonite color only in connection with the particularities of its composition and structure without the use of specialized spectrometric methods [47].

Over the course of recent decades, certain researchers (M.G. Isakov, V.P. Kutz et al.) have continued to ascribe to rubidium the role of a color-forming element, supporting this hypothesis only with data from chemical-spectral analysis of amazonite. Other researchers (E.N. Eliseev, K.K. Zhirov, M.N. Ostrooumov), on the contrary, have brought attention to amazonites occurring with variable rubidium contents, while noting common potassium feldspars occurring with rubidium content greater than that found in amazonite. These facts have eventually enabled researchers to reject the significance of rubidium as a chromophore (G.S. Pliusnin et al.).

Taking the place of the rubidium hypothesis, works are beginning to appear on the role of iron in amazonite color (E.N. Eliseev et al.). In chemical analyses, amazonite is consistently found to contain both of the valent forms of iron, which, as is known, participate directly in the color of many minerals, including feldspars (M.N. Ostrooumov). Precisely for this reason, as well as following from the results of experimental measurement of amazonite color (examined in detail below), it was presumed that the various forms of iron exerted an active influence on amazonite color of K-feldspars. However, still other models have been developed that reject a color role for this element (K.K. Zhirov et al., G.S. Pliusnin, S. Taylor) [47]. S. Taylor and collaborators, not finding significant differences among the compositions of variously colored microclines (containing an entire range of elements, including iron), refute in general the opinion that amazonite color is correlated with its chemical composition. Worth noting as well is G.S. Pliusnin's categorical exclusion based on the measurement of magnetic sensitivity of several variously colored amazonite specimens, conducted before and after heating to 900 °C: according to electron paramagnetic resonance (EPR) data, Fe^{2+} is absent in amazonite; iron is present in amazonite only in its trivalent state.

As can be seen, none of the aforementioned hypotheses have provided a complete and satisfactory explanation for the reason behind amazonite color. This situation has led to the appearance of a range of proposals for the nature of amazonite color.

In 1954 I. Oftedal, studying the geochemistry of the amazonitic pegmatites of southern Norway, brought attention to the ingression of a notable quantity of fluorine during the amazonitization process, which allowed him to propose a correlation of amazonite color with an impurity of this element; fluorine ions replace oxygen ions, thereby creating an unstable distribution of charges and, as a result, color [32]. In 1959, K. Pshibram, using data on the luminescence and radioactive irradiation of heated amazonites as well as citing their chemical analyses, proposed that amazonite color should be considered radiogenic and driven by manganese impurity.

In 1959, K.K. Zhirov with collaborators first accented attention on the lead content of amazonites being two to three times greater than that of associated potassium feldspars of common color. They established that the lead content in amazonites from various source deposits varies from 0.008% to 0.082% (in several cases up to 0.1%), with the more intensively colored varieties corresponding to higher concentrations of lead.

Works on the minerals' structural state comprise another tendency of amazonite research. The authors of several studies (F.T. Sandford, T. Hedvall), not finding evidence of the participation of elements impurity in the amazonite color process, attempted to explain the color as solely the influence of the fine structural imperfections (defects) of a microcline; the character of the latter in these works was not examined. The phenomenon of defects was linked to the processes of ionic substitution. By the beginning of the 1960s, S.M. Stishov's theory of the structural defect nature of amazonite color became fairly widespread in the literature, providing a concrete model of a color center based on an impurity of lead ions.

The detailed structural characteristics of amazonite were first reported in 1964 in the work of L.A. Ratiev and Kh.N. Puliev, who used methods of X-ray structural analysis and infrared spectroscopy to prove that it belonged to the more highly ordered phase of potassium feldspar—the maximum microcline. These circumstances, as well as the known facts about the thermal decoloration of amazonite and the decrease of Al-Si order of a microcline under heating enabled the authors to posit that the maximum degree of structural order of potassium feldspar should be considered one of the necessary conditions for the phenomenon of amazonite color.

In 1967, A.N. Bugaets obtained important data on the structure of amazonites from the amazonitic granite of Kazakhstan [2]. He established that the majority of amazonites from granites were associated with the intermediary triclinic orthoclase or (a minority of them) pseudo-twinned microcline, and that amazonites from vein derivative granites were associated with the maximum microcline. Bugaets emphasizes that an essential role in the phenomenon of amazonite color of potassium feldspars is played by the order–disorder of the structure and the absence of albite components. According to the author's observations, amazonite color occurs only in a high degree of ordered potassium feldspars that are subjected to disintegration, segregation, and purging of albitic components.

Regarding amazonites from granites and associated pegmatoid and hydrothermal veins, L.G. Feldman with collaborators [2] observed the gradual (parallel with or even outpacing an increase in intensity of amazonite color) structural order of potassium feldspar, manifested in the occurrence in the grains of the twinned lattice (initially unclear, then increasingly pronounced) and the corresponding increase of the angle of the optical axis and X-ray triclinity. It is worth noting that these presentations in essence support on a new level the forgotten observations of A. Des Cloizeaux (see Section 1.1). The authors arrived at the conclusion that the structural order process of potassium feldspars, including foremost in the order of Al and Si, is accompanied likewise by the order of K and the changing of the interatomic spacing of K-O. As a result of the weakening of bonds in the nonequivalent structural positions, a concentration occurs of those element impurities that form the defect centers of amazonite color. In this work, definitive progress was made toward harmonizing the previously disparate data on the chemical composition (Pb, Rb, OH, etc.) and structural state of potassium feldspars as it relates to amazonite color.

In 1968, B.M. Shmakin, researching amazonites from a range of pegmatitic deposits, considered two factors as necessary conditions for the phenomenon of amazonite color in potassium feldspars: high structural order and elevated element impurity content. According to Shmakin, the development of amazonite color is primarily a consequence of the order of the mineral's crystal structure; its decoloration under heating is caused by the disorder of its structure from triclinic to monoclinic; and the subsequent irradiation to X-rays enables the reverse transition of feldspar to the triclinic modification and the recovery of amazonite color. Shmakin noted as well that the twinned lattice characteristic of amazonites disappears under heating and reappears under exposure to harsh radiation. Unfortunately, these findings were not supported by quantitative measurements of the color and structural state of amazonites after heating and subsequent X-ray radiation, rendering this a purely theoretical conclusion.

At present, there is little doubt that amazonite color depends on its structural state. The aforementioned general particularity of this dependence—the increase of the intensity of amazonite color in proportion to the increase of the structural order—should be emphasized. As a natural consequence, amazonite is understood as an entire range of structural modifications of potassium feldspar, in which the maximally ordered phase (maximum microcline) is considered only an extreme, more intensive component of color, despite the confirmation of several researchers that amazonite always presents as a maximum microcline.

We move now to the characteristics of amazonite color research conducted on the basis of various spectrometric methods and laboratory experiments.

In 1949, E.N. Eliseev carried out a specialized study of the possible participation of impurity of d-elements in amazonite color. Drawing on research of spectra of amazonite transmission, the author produced findings on a connection between the absorption band in the range of 600 to 650 nm and ions of iron (II) oxide, and on the occurrence of amazonite decoloration under heat and the partial recovery of color after irradiation, explained by the oxidation-reduction process $Fe^{2+} \rightarrow Fe^{3+}$.

As one of the essential arguments supporting his proposal, E.N. Eliseev described the occurrence in specimens after heating of secondary yellow or reddish-brown color, formally corresponding in tonality and in character to the optical absorption of the color of feldspars containing the unstructured form of iron. These observations subsequently were supported more than once by the experiments of other researchers (K.K. Zhirov et al., M.N. Ostrooumov) who noted, however, that not all amazonite specimens after heating acquired yellow or reddish-brown hues; several became white under thermal decoloration.

Subsequent criticism of this hypothesis was based primarily on two points: (1) the low iron content, especially Fe^{2+}, in microclines, including amazonites; and (2) the absence of a correlation between the intensity of amazonite color and its iron content. These arguments at first glance are quite solid: total content of iron impurity in microclines ($Fe_2O_3 + FeO$) in rare cases reaches 1%, and in unique specimens of the yellow ferro-orthoclase in pegmatites from Madagascar the concentration of Fe_2O_3 does not exceed 1%. In amazonites the maximum Fe_2O_3 content consists of 0.5% and FeO of 0.1%, that is, it fluctuates within the margin of error of traditional methods of chemical separation of oxides of iron of different valences. Naturally, when relying solely on the results of chemical analyses of amazonites, it is implausible to speak of any correlation, although several qualitative dependences deserve attention. Thus M.N. Ostrooumov and others have established that the general iron content in amazonites and especially the relationship of FeO to Fe_2O_3 impacts the position of long-wavelength (625–720 nm) absorption bands of amazonite and its intensity—in blue amazonites the significance of this ratio is minimal (0.17); while in green varieties, it reaches 1.33.

There is sufficient support for the hypothesis on a connection between amazonite color and the ingression into its structure of lead ions, first expressed and experimentally supported in 1959 by K.K. Zhirov and collaborators and developed subsequently by A.S. Marfunin [30]. Shortly afterward appeared works proving the validity of the new hypothesis; as a rule, they cite the results of rigorous chemical analyses that provide evidence that lead concentration in amazonites is elevated by comparison with that found in common K-feldspars (V.P. Kuts, H. Makart, A. Preisinger).

At the same time, this hypothesis, in the way it was formulated by its authors, likewise did not provide an exhaustive explanation for amazonite color, but served as the basis for the creation of new, more developed variants of the lead hypothesis. The authors of one of these [27], noting a direct correlation between the intensity of amazonite color and its lead content, propose that the alteration of color depends, possibly, on the electron transition $Pb^{2+} + Fe^{3+} \rightarrow Pb^{3+} + Fe^{2+}$. In essence, this was the first serious attempt to analyze the multicausality of the phenomenon of amazonite color. In another variant [2], amazonitization was linked to the displacement of potassium by rubidium and lead, and of oxygen by fluorine and hydroxyl (OH group), which causes the unbalanced structural state of feldspar and the inception of defect centers of amazonite color.

An extreme point of view was expressed in 1971 by F. Čech, Z. Mizar, and P. Povondra, in their detailed research of green, lead-containing (up to 1.19% PbO) orthoclases from the deposit at Broken Hill, Australia. They proposed that the presence of lead was the reason for the green color not only in microclines, but in all other feldspars. According to this reasoning, they raised the question of reclassifying all feldspars with green color under the name amazonite.

The data obtained by G.S. Pliusnin was employed as evidence for the correlation of amazonite color to the content not only of lead in the position of potassium but also of water or the OH group in the structure of amazonite [1]. Examining the hydroxyl group in its capacity as a charge compensator under the heterovalent isomorphism $K^+ \rightarrow Pb^{2+}$, he considered that its ingression into the structure of feldspar results in the displacement of the absorption band characteristic for amazonite and consequently produces its green or blue color. Also of note are Pliusnin's highly interesting experiments with heating amazonite in various media. According to his data, which confirm the earlier experiments of I. Oftedal, amazonite fully decolorizes under heating in air over the course of more than 10 hours at temperatures of 250 to 270 °C. Under heating of such a specimen from the autoclave with distilled water at a temperature of 300 °C and applying pressure of 7 MPa over the course of 48 h, green amazonite does not lose color, but it acquires a distinct blue tint.

In the opinion of the author of the present work, this experience provides unambiguous evidence that amazonite's basic absorption band of 630 to 640 nm is not connected directly with electrons or electron hole sites with thermal stability of 300 °C and lower. However, an insignificant decrease in the intensity of the maximum of reflection and a change of the tonality of color could be the consequence of the destruction of another color center (for the given specimen) that caused the appearance of green hues. On the other hand, evidently, a water medium inhibits the destruction of amazonite color and possibly serves as one of the reasons for the development of amazonitization.

Crucial significance was assigned to the structural forms of water in the process of amazonite color by A. Hofmeister and G. Rossman [28], who proposed that in the process of γ-radiation molecules of water dissociate into the formation of $H°$ and $OH°$. These radicals, which represent effective electron traps, play an important role in the formation of electron-hole color centers (for example, type O^-), and likewise in the processes of valence conversion of impurity of lead ions in the structure of amazonite.

Further development of the lead hypothesis was connected with the first (EPR) research of amazonites, performed in 1970 by A.S. Marfunin and L.V. Bershov. They established that the electron center $Pb^+(Pb^{2+}+e^-)$, observable only in amazonites, under heating disappears simultaneously with color; thus, this electron center may be considered the reason for color.

In 1984, A.N. Platonov and collaborators conducted research of the spectra of absorption and luminescence as well as the thermoluminescence and thermodecoloration of amazonites, which provided further proof of the participation of lead in color of this mineral [20]. In the spectra of amazonite, they detected three characteristic absorption bands with maximums of 255 nm (greatest intensity), 390 nm, and 625 nm, the latter two of which explain amazonite color. According to the author, the UV-range band, the intensity of which was established to have direct dependence on lead concentration, is caused by the electron transition $^1S_0 \rightarrow {}^3P_1$ in ions of Pb^{2+}. As distinct from the spectra of luminescence of common feldspars with dual-band radiation (470 and 730 nm), observed in the spectra of luminescence of amazonites was a band of around 285 nm, attributed to the transition $^3P_1 \rightarrow {}^1S_0$ in Pb^{2+}. The absence in the 285-nm band of thermodecoloration maximums within the range of amazonite thermo-decoloration allowed these researchers to propose the existence in feldspars of two structurally nonequivalent positions of Pb^{2+}, which are defined by the character of local compensation. A large part of these ions are isomorphically displaced potassium ions that possess a stable bivalent state and occur only in the role of centers of luminosity; Pb^{2+} ions, not having charge compensation, were able under activation to change valences and, apparently, to play an essential role in the formation of color centers in natural amazonites.

It is worth noting that A. Hofmeister and G. Rossman, likewise, tend to favor this hypothesis [28], explaining amazonite color in terms of the electron transitions in Pb^{3+} or Pb^+ ions that form as a result of the capture of an electron hole or an electron by impurity of Pb^{2+} ions under the action of natural radiation. The authors, positing an analogy between the optical spectra of amazonite and that of $KCl:Tl^{2+}$, are inclined to consider Pb^{3+} as the basic chromophore in amazonites.

Finally, the most recently conducted works surveyed here [29,48] suggest the presence in amazonites of other specific centers of lead. One work [48] proposes that the term amazonite denote a blue-green regular microcline with mixed valence pairs of lead $(Pb_A–Pb_B)^{3+}$ acting in the capacity of a chromophore. On the other hand, A. Julg [29], producing a theoretical study of the absorption spectra of aliovalent lead (Pb^+ and Pb^{3+}) ions in a microcline, proposes that precisely these are responsible for a specific color of amazonite; in other words, Julg supports the earlier hypothesis of A. Hofmeister and G. Rossman [28].

From this discussion of the various hypotheses for the reason for amazonite color in feldspars, it can be stated with certainty that amazonite color has a complex nature, that is, it is caused by the aggregate of several mutually inclusive factors, each of them necessary components in the amazonitization process.

In the analysis of the literature dedicated to amazonite color, attention is called to the following circumstances. First, there is the insufficient state of the knowledge on color itself—while the visually observed variety of the colors of amazonite is widely acknowledged, the spectrometric works so far conducted do not reflect the actual picture of the mutability of this property. Second, in these works, the comparison of the optical spectra of amazonites containing some or other element impurities usually takes into account only one parameter of the spectra—the intensity of the absorption band that explains the specific color—and does not account for the shifting of this band along the length of the waves. Finally, we note that among the multitude of hypotheses purporting to explain the reason for amazonite color, still there are none that would satisfactorily treat the complex nature of amazonite color of feldspar.

The genetic aspects of the amazonite problem after recent discovery and detailed study of large masses of amazonitic granites have once more captured the attention of a range of geologists. Researchers who study these rocks and the associated pegmatoid and hydrothermal formations have advanced a hypothesis for the dual nature of amazonite, that is, the possibility of the formation of this mineral by two means: primarily as a result of the amazonitization process of feldspars, and to a lesser degree by the direct crystallization of amazonite in hydrothermal veins [2].

Alongside such conceptualizations, there continue to exist still other hypotheses that deal singularly with the genesis of amazonite and in essence correspond with those of A.E. Fersman and A.N. Zavaritsky. Among the followers of Fersman (V.I. Kovalenko, N.I. Kovalenko et al.), the hypothesis remains accepted that amazonite developed in different genetic types of rock directly crystallized from igneous melt rich in volatile components and rare elements. Adherents of the second tendency, who connect the genesis of amazonite to post-igneous processes, are concerned principally with the question of what is meant by the amazonitization process. One of these studies (A.A. Beus et al.) classifies it as the metasomatic conversion of the previously recrystallized granites, proposing that the occurrence of amazonite is connected to subsequent microclinization (K-feldsparization), while others view it as the special colorization process of the previously formed potassium feldspars of the granites that occur following the processes of albitization and greisenization [1].

In conclusion, it should be noted that in the literature, the practical question of the prospecting significance of amazonite has not received sufficient elucidation.

In essence, only one work [2] has performed a serious analysis of the link between the ore content of granites and the development of the amazonitization process in the latter. In view of the presence of amazonite as a prospecting indicator for the detection of tantalum minerals in granites, the authors of that work consider that the absence of amazonitization in granite plutons cannot serve as an indicator for a lack of ore. At the same time, they point to amazonite as a beneficial indicator of the occurrence in granites of the greisenization and albitization processes and their associated ores. Such a contradictory evaluation of the prospecting significance of amazonite is elicited probably by the fact that the concrete features of the crystal chemistry and properties of this potassium feldspar are nonuniform in granite plutons of varying amounts of ore. These features in every case must be taken into account; in other words, for every case, it is necessary to conduct a complex of detailed investigations of the composition, structure, color, luminescence, and other properties of this mineral.

Unfortunately, even the proponents of the practical prospecting-evaluating use of amazonite have not put forth serious arguments and justifications in support of this important capacity. Finally, it is worth noting the complete absence in the geological and mineralogical literature of any kind of data on the typomorphic significance of amazonite.

Concluding this brief survey of the research on the crystal chemistry, color, genesis, and prospecting significance of amazonite, we emphasize that the weak point of almost all experimental work has been the insufficient consideration or practically complete ignoring of the geological setting of this mineral. On the other hand, in works of geological, mineralogical, and to some degree genetic character, there is a clear deficit of experimental data. In the majority of these (see Chapter 2), amazonite, as a rule, is mentioned only in connection with various mineralogical and geochemical specializations of rock formations and ores.

CHAPTER 2

Geological Setting of Amazonite

2.1 GEOGRAPHY OF AMAZONITE

The numerous guide, reference, and textbooks on the mineralogy of the feldspar group usually identify an entire series of varieties, but the degree of their distribution is not evaluated. Probably for this reason, amazonite was considered for a fairly long time a relatively rare mineral, and the well-known Il'menskie mines in the southern Urals were recognized as the largest amazonite deposit.

By now, amazonite has been discovered in many districts in Russia and in other countries, and it is no longer seen as a rarity in the mineral world as it gradually becomes known in a vast number of finds. However, it is very difficult to measure the significance of these finds, insofar as the literature has not always provided information on the scale and the intensity of the development in different genetic types of rock of the amazonite formation process. Indeed, the separate, small in size, pale-colored crystals of the amazonitic feldspar found in the pegmatites of India and the Pamir Mountains cannot be compared with the large blocks (up to 2 m in diameter) of amazonite with the bright, saturated coloration found in the pegmatites of the deposits of the Kola Peninsula, the Baikal region, and Tuva, where this mineral can comprise up to 90% of the vein bodies.

To cite one still more contrasting example, in subvolcanic rocks (ongonites) amazonite has so far been recorded only in a few dike bodies in a type of very fine (0.2–0.7 mm) and weakly colored rare porphyritic crystals (porphyroblasts),

Amazonite: Mineralogy, Crystal Chemistry, and Typomorphism. http://dx.doi.org/10.1016/B978-0-12-803721-8.00002-0

occurrences that can be considered exotic. On the other hand, plutons of amazonitic granites with outcroppings of significant area (up to several, sometimes dozens of square kilometers) are layered with a huge number of relatively small crystals of that feldspar, the specific coloration of which may vary as much by tonality as by intensity within fairly wide margins. That is why cases are known where small or rare individual amazonites with pale colors developed in various rocks simply do not attract attention and are not recorded.

Regarding the stated, undoubted wide distribution of amazonite, in the subsequent description attention is devoted principally to its predominant deposits and occurrences: first, pegmatite bodies, which have been developed or might come under development for the purpose of obtaining of ornamental stone; and second, those amazonitic granites (and metasomatites) in which amazonite occurs as the main rock-forming or as a secondary mineral and is associated with accessory rare-metal mineralization.

A brief review is here presented of the geography of the main deposits and occurrences of amazonite in Russia and several countries of the former USSR (Commonwealth of Independent States, CIS).

Urals. Amazonite was first discovered in bedrock deposits at the end of the eighteenth century in the pegmatite veins of the Il'menskie Mountains [6,18]. In the numerous classifications of granite-pegmatites of this region, almost all researchers have distinguished two principal types of pegmatite veins: amazonitic and quartz-feldspathic (without amazonite). For the first type, an accessory rare-metal and ornamental gemstone mineralization has always been considered characteristic.

At the end of the 1950s, in a geological situation analogous to the Il'menskie mines, M.G. Isakov discovered amazonite-containing pegmatites in the Vishnevye Mountains and the Uvil'dy alkali band. Moreover, according to the data of A.I. Sherstiuk, pegmatites with amazonite are noted in the boundaries of the Murzinka granite pluton.

Kola Peninsula. Amazonite-containing pegmatites were discovered by O.A. Vorob'eva in 1928, but intensive study of them began only from the end of the 1950s, primarily as a source of beautiful ornamental material (I.V. Bel'kov, E.Ya. Kievlenko, A.Ya. Lunts et al.). To date, several fields of development of amazonitic pegmatites have been established here, of which the largest and most typical in geological structure is the western Keivy. Other, lesser fields of pegmatites with amazonite have been discovered in the upper and middle Ponaia River, the upper Strel'na River, near the lakes Kedik'iavr, Seiiavr, Kanozero, and Chunatyndra, and in other districts.

Karelia. In northern Karelia (in the Loukhi District), geologists of the northwestern geology association (Sevzapgeologia) have discovered highly unique amazonite-containing pegmatite: the Pirtim deposit and a series of others [1]. The study of their geological setting, internal structure, mineralogy, geochemistry, and other particularities enables a hypothesis on the prospects of this district with regard to its occurrence of pegmatite deposits with tantalum-niobium mineralization and raw gemstones.

Eastern Siberia. The presence of amazonite in the composition of pegmatites of the Baikal region was established by the 1940s. Numerous pegmatitic fields with amazonite and accessory Ta-Nb oxide minerals were found subsequently by M.M. Manuilova and other researchers in the northern Baikal region. Pegmatite deposits with amazonite, according to the data of B.M. Shmakin and P.V. Kalinin et al., were also widely developed along the north-western shore of Baikal (between the Ust'-Unga Gulf and the Ulan-Nur cape) and in the Sliudiansk District.

In eastern Saian (Buriatiia), amazonitic pegmatites and granites were described by V.A. Dvorkin-Samarskii. In Tuva in the upper reaches of the rivers Kachik, Chakhyrtoi, and Baiangol, Iu.L. Kapustin encountered amazonite-containing pegmatites, and subsequently ongonites with amazonite.

In eastern Transbaikal, one of the plutons of amazonitic granites was known and was under development for rare metals already in the 1930s; in the late 1950s to early 1960s, a province of such granites significant in terms of horizon was discovered; according to the data of a range of researchers (A.A. Beus, L.G. Fel'dman et al.), a series of intrusive bodies among these granites was recognized as potentially rare metal bearing.

CIS Countries. Ukraine. In the Azov region of the Ukrainian crystal shield, amazonite has been uncovered in pegmatites genetically connected with Proterozoic granitoids, in the opinion of V.P. Kuts. Separate finds of amazonite are known in the pegmatoid build-ups of the Kamennomogil'sk and Dubovskii granite plutons, and in the pegmatites at the villages Staraia Ignati'evka and Ekaterinovka. A small occurrence of this mineral was established by N.A. Bespal'ko in Zhytomyr Oblast in the interfluvial region of Perga and Uborta; it coincides with the horizon of development of Perga granites.

Uzbekistan. Amazonite is known in the pegmatite veins of the Oiganskii field, deposited in endo- and exocontacts of the Barkrak granitoid pluton. Furthermore, amazonite has been noted by I.Kh. Khamrabaev in pegmatites situated to the north of the Zirabulak intrusive in Darantut sai.

Tajikistan. According to the data of E.A. Dmitriev et al., amazonite is found in the southern Pamir Mountains in pegmatites of the Mazar pluton and in granites of the ore occurrences of Pisodu, in the eastern Pamirs in pegmatites with accessory topaz-beryl mineralization, and in the south-western Pamirs in pegmatites of the veins of the Davlakh complex of granites and in pegmatites of the Raumid granite pluton. Separate occurrences of amazonite have been identified by V.D. Dusmatov in pegmatites in the southern slope of the Gissar Range, including in the vicinity of Dushanbe, and in central Tajikistan.

Kazakhstan. Among the rare-metallic albitic granites of Kazakhstan, many researchers have distinguished an amazonite-albitic mineral type with a highly characteristic geochemical specialization of rare elements. Amazonite-albite granites, according to G.P. Lugovskii et al., have been mapped in more than two dozen granite plutons (Maikul', Inyl'chek, Khorgos, and elsewhere). Analogous intrusive bodies have been identified also in **Kyrgyzstan** in the northern Tian Shan (Kurmenty and elsewhere).

In recent decades in many districts of Russia and certain CIS countries, there have been discoveries of new deposits and occurrences of amazonitic pegmatites (Baikal region, Kolyma, Yakutiia, Chukotka, the Far East, and Central Asia), granites (eastern Saiian, Kazakhstan, and Kyrgyzstan), and ongonites (eastern Siberia).

We now turn to the geography of the principal deposits and occurrences of amazonite outside Russia.

Asia. By the late 1950s in the territory of Mongolia, V.I. Kovalenko and other researchers discovered and described numerous plutons of amazonitic granites with rare-metal mineralization. One such pluton contained pegmatite veins with amazonite, topaz, and crystal-bearing mineralization. According to Iu.O. Lipovskii, amazonitic granite from the Abdar pluton is used in construction as a facing material and for the fabrication of bricolage. N.I. and V.I. Kovalenko were the first to discover in Mongolia ongonites—subvolcanic analogs of rare-metallic granites sometimes containing amazonite.

Amazonitic granites with rare-metallic minerals are known also in China.

India. In the states of Kashmir and Andhra-Pradesh (the veins of Shankar and Pilimitt in the vicinity of Nellore), rare-metallic and muscovitic pegmatites with amazonite are intensively explored. Amazonite is noted also in the cassiterite-containing pegmatites in association with lepidolite and a range of rare-metallic minerals in the Bastar pegmatite field (Madhya Pradesh state). Amazonite in mica pegmatite in the district of the city Kodarma (Jharkland state) was described in 1952 by E. Spenser.

According to B.M. Shmakin, bluish amazonite in association with pink mica, biotite, albite, schorl, polychromatic tourmaline, and beryl occur in the pegmatite deposit of polychromatic tourmaline at Kh'iakul in eastern Nepal.

L.N. Rossovskii (personal communication) has observed amazonite in the spodumene pegmatites of Afghanistan. There are data on a find of amazonite in Saudi Arabia in the vicinity of the Red Sea.

Europe. Pegmatite deposits and occurrences with amazonite are known in Austria (Pack, Styria), Bulgaria (village Kesten, Smolyan district, in the central Rhodopes), Norway (Søndeled; Terdal; Telemark; Iveland; and elsewhere), Sweden (Uttögrube; Utö; Ytterby; Skantorp; and elsewhere), Poland (Strzegom-Sobótka), Czech Republic (city Zdan; Hajany near Blatná), Macedonia (Chanishte; Vitolište; and Prilep), and Germany (Bodenmais, Bavaria). A range of works by Russian and other researchers contain contradictory and insufficiently proven data on finds of amazonite in Italy (Baveno, Piedmont), Switzerland, France,

Spain (Galicia), Great Britain (Meldon, Devon county; Scotland), and Finland (Tammela).

Africa. It appears that this continent is home to the earliest finds of amazonite: in the Eastern and Nubian deserts, as well as in the upper Nile (contemporary Egypt, Sudan, and Ethiopia). In connection with the intensification of geological works in recent decades, numerous deposits of granites and pegmatites with amazonite have been discovered in Egypt (Sinai Peninsula and elsewhere), Algeria (Hoggar in the vicinity of the Mali border and elsewhere), northern Nigeria (Jos Plateau), Burundi, Zimbabwe, Cameroon, Kenya, Mozambique, and Rwanda.

The literature is scant of data regarding deposits in southern Africa (Republic of South Africa and Namibia). A significant distribution has been proven for amazonite in the pegmatites of Madagascar, as described by A. Lacroix in 1922.

Australia. Green feldspar, long known in the deposit of Broken Hill (New South Wales), was recognized as amazonite relatively recently by F. Čech et al. L. Arens in one of his works referred to amazonite from Australia (no location was given). The fairly comprehensive summary monograph *Mineral resources of Australia and Papua New Guinea* (K. Night, ed., 1980) notes a explored pegmatite deposit of amazonite in the vicinity of Port Lincoln (South Australia).

North America. In the USA, several pegmatite deposits in the states of Virginia (Amelia Courthouse) and Colorado (Pikes Peak) have been explored particularly for amazonite. Finds of amazonite in pegmatites are noted in Pennsylvania (Delaware County), North Carolina (Ray Mica Mine), South Dakota (Black Hills, Glendale), Massachusetts (Rockport, Gloucester), and California (Pala).

Relatively recently, amazonitic granites have been discovered and described in Mexico (Chihuahua state) [40]. Large pegmatite deposits with amazonite are known in Canada (Ontario, Québec).

South America. Several amazonite occurrences have been noted in Brazil (but not in the Amazon basin) in the state of Minas Gerais (Cuieté, Conselheiro Pena; Itabira; Conceição; and elsewhere), which has long been considered one of the largest regions of concentration of pegmatite deposits with gemstone mineralization [7].

In conclusion, we supply a brief inventory of the principal pegmatite and granite deposits associated with the largest deposits of amazonite.

Russia and CIS Countries
Pegmatite Fields and Deposits:
1. Western Keivy pegmatite field (Kola Peninsula)
2. Il'menskii Mountains (southern Urals)
3. Ulan-Nurskoe
Pegmatite Occurrences:
4. Naryn-Kuntinskoe, Abchadskoe, Ainskoe, Samsal'skoe, Zashikhinskoe, Baian-Kol'skoe, Sliudianskoe, Poperechenskoe, Turakinskoe, Usmunskoe (eastern Siberia)

5. Barkrakskoe, Darantutskoe, Rauid-Darinskoe, Turpinskoe, Iarzyvskoe, Rangkul'skoe, Mazarskoe (Central Asia: Uzbekistan and Tajikistan)

Deposits and Occurrences in Granites:
6. Bazardarinskoe, Karaobinskoe, Maikul'skoe, Kengkiikskoe, Katbasarskoe, Dogolanskoe, Akteisiauskoe, Kyngyrzhal'skoe, Narymskoe, Khorgosskoe, Chizhinskoe, Inyl'chekskoe, Oisazsckoe, Kurmentin, Koturginskoe, Taldynskoe, Tuiuksuiskoe, Aksaiskoe, Tastynskoe, Karagaily-Aktasskoe, Dzhylysuiskoe, Keregetashskoe, Tonkskoe, Raumidarinskoe (Central Asia and Kazakhstan)
7. Khoroiskok, Saivonskoe, Bitu-Dzhidinskoe, Oimurskoe, Bezymianskoe, Kutimskoe, Nerugandinskoe, Kharagul'skoe, Shagaite-Gol'skoe, Khukhu-Chelotuiskoe, Khangilai-Shilinskoe, Gornachikhinskoe, and elsewhere (eastern Siberia)

Others Countries
Deposits and Occurrences in Granites:
1. Abdar, Baga-Gazryn, Iudygin, Zhanchivlan, Borun-Tsogtin, Bural-Khangai (Mongolia)

Deposits and Occurrences in Pegmatites:
2. Ambokhitravorano, Sakhavorona, Ampangabe, Andina, Andriamena, Makhabu, Soarano, Andibakeli, Makharitra, Sakhafurana (Madagascar); Shimanda (Zimbambwe)
3. Budkoks, Renfrew, Perry Sound (Ontario, Canada), Bouchette-Hall, Ville-nieve, Leduc (Québec, Canada)
4. Amelia, Rutherford, Morefield (Virginia), Pikes Peak, Crystal Peak, Florissant, and Cameron Cone (Colorado), Rockport (Massachusetts), as well as occurrences in New Jersey, New Mexico, New Hampshire, North Carolina, South Dakota, Pennsylvania, and California
5. San Miguel de Pirasicaba, Joaoxuma, Ferros, Itabira, Concepção (Brazil)

We enumerate further the remaining locations known to author et al. [47] in Russia of amazonite finds of mineralogical interest:

1. Kola Peninsula: Ura-Guba, Shongui, Parus city, Rovgora
2. Leningrad Oblast', Vyborg District: Vozrozhdenie
3. Central Urals: Shartash; northern Ural
4. Altai: Kalbinskii Ridge
5. Northern Baikal region: Davan, northern Muiskii ridge
6. Southern Baikal region: Komar ridge, Bol'shaia Bystraia River basin
7. Iakutiia: Tommot
8. Kolyma: Chaunskaia Bay (northern pluton)
9. Transbaikal: Adun-Cholon, Bukuka
10. Khabarovsk Krai: Dusse-Alin', Bureinskii Ridge
11. Primorskii Krai: Tigrinoe (central Cikhote-Alin')

2.2 AMAZONITE'S GEOLOGICAL SETTING AND SPATIAL-TEMPORAL DISTRIBUTION PATTERNS IN GRANITOID FORMATIONS

Examining the particularities of the distribution of the principal amazonite deposits (Fig. 2.1), above all it should be noted that the majority of pegmatite deposits are located on the borders, rarely within the boundaries of the Precambrian platforms (Siberian, North American, Afro-Arabian, and Australian) and shields (Baltic, Ukrainian, Greenland, and Indian). An altogether small part of them is known in fold belts: Appalachian (Caledonides), Ural (Hercynides), Pacific Ocean (Cimmerides), and Mediterranean and Pamir (Alpides). On the contrary, the plutons of amazonitic granites and their metasomatites occur primarily in the orogens of Cimmerides (Transbaikal, Russia; Mongolia) and to a lesser degree Caledonian (eastern Saiian, Kazakhstan) and Hercynian (Tian Shan) ages.

The review of factual material on amazonite brings to light the absence among the multitude of studies of a general systematization of all known data on the geological setting of amazonite finds. In part, this situation might be explained by the fact that for a long period, a classification system has not been proposed for this purpose.

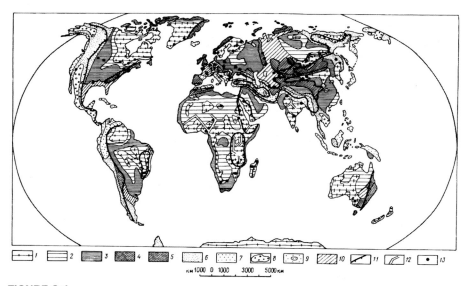

FIGURE 2.1
Distribution of the principal deposits and occurrences of amazonite (tectonic base according to V.V. Belousov with simplifications): (1) shields; (2) slopes of shields and anticlises of ancient platforms; (3) sineclises of ancient platforms; (4–7) geosynclinals (4: Caledonian, 5: Hercynian, 6: Cimmerian, 7: Alpine and Pacific); (8–9) central plutons in geosynclinals (8: Cimmerian, 9: Alpine and Pacific); (10) Epi-Paleozoic platforms with sedimentary cover; (11) boundaries of neo-tectonic orogenic and rift belts; (12) large trough fault; (13) deposits and occurrences of amazonite.

The geological setting of amazonite is taken as the basis for the classification of amazonite-containing formations described below; geological setting refers to amazonite's connection with defined granitoid or formations, its occurrence in defined genetic types of rock, and its distribution in various geological epochs, among other things [1]. The materials laid out in this section are based primarily on the knowledge of a range of amazonite deposits and occurrences in Russia and in countries of the near and far abroad, as well as on the analysis of numerous datasets from the literature.

As was shown in the first edition of the monograph [1], arising in the course of the inversion (according to T.N. Spizharskii) of the tectonic–magmatic cycle are the following granitoid formations: granodioritic (and the formations isomorphic to it), granitic, leucogranite-alaskite, subalkaline-leucogranitic (or fluorine-lithium granites), and alkaline granitic—all markedly dissimilar in geochemical specialization (Table 2.1) and potential ore-bearing mineralization. Plutons of granodioritic and granitic formations are not associated with rare-metal deposits. Plutons of leucogranite-alaskites produce mutually exclusive types of deposits: miarolitic pegmatites; rare-metallic (non-lithium) pegmatites with beryllium and tantalum-niobium mineralization; quartz-vein-bearing greisenic tin, tungsten, molybdenum, bismuth, and beryllium; albitic-greisen-bearing tin, and niobium, and more rarely tungsten, beryllium, and tantalum. Some minor intrusions of subalkaline granites, localized in narrow-lined zones, generate in domes and apophyses rare-metal deposits and their associated endo- and exocontact quartz-vein-greisenic deposits, which include tin, thallium, lithium, beryllium, fluorine, and more rarely rubidium, cesium, and tungsten; they can also occur in a type of belt of rare-metal lithium pegmatites containing lithium, rubidium, cesium, beryllium, and tantalum-niobium mineralization. Domes and apophyses of plutons of alkaline granites tending to have the same geologic structural specificity coincide with other rare-metal deposits and their associated endo- and exocontact metasomatites with niobium, rare-earth elements, zirconium, fluorine, beryllium, and thorium (and sometimes lead); occasionally, these plutons produce rare-metal and rare-earth alkaline-granitic pegmatites with niobium, zirconium, rare-earth elements, yttrium, and beryllium.

The overwhelming majority of granitoid plutons and ore clusters is comprised of the genetic types of several formations (telescoping), whereby in a concrete cluster each of the rare-metal-bearing formations presents a single formational subtype.

Occurrences of amazonite are associated with the defined genetic types of rocks of three granitic formations: leucogranite-alaskite, subalkaline-leucogranite, and alkaline-granite. Amazonite is found in granites, aplites, pegmatites, metasomatites, feldspar-quartz veins, as well as in subvolcanic rocks, that is, ongonites. Table 2.2 illustrates the distribution of amazonite in various genetic types of rocks and formations and reflects the known or proposed age and formation depth of concrete amazonite-containing formations and their characteristic mineral associations.

Table 2.1 Average Contents (ppm) of Elements-Impurities in Rocks of Granitoid Formations

Element	Granitoid formation type							Average in granitoids (according to A.I. Vinogradov)
	Gabbro-diorite	Diorite plagiogranite	Diorite granodiorite	Granite	Leuco granite-alaskitic	Subalkaline-leucogranite	Alkaline granite	
Cr	190	60	40	20	9	4	5	25
V	220	95	70	30	15	5	3	40
Co	40	15	10	7	2.5	2	1	5
Ni	80	50	15	10	4	2	1	8
Cu	44	32	25	18	12	10	5	20
F	200	300	600	700	1400	3800	1600	800
B	10	13	12	18	35	12	7	15
Rb	45	75	110	190	290	600	260	200
Li	12	18	25	40	75	220	90	40
Cs	0.1	0.4	1.3	3.2	8	25	9	5
Tl	-	0.1	0.5	0.8	1.3	3	1.3	1.5
Be	0.7	1.5	2	3	6	9	5	5.5
Sr	400	600	400	300	100	20	30	300
Ba	350	550	850	700	250	50	75	830
Sn	2.1	3	3.9	6	9	20	7.6	3
W	0.1	0.4	0.8	1.9	3	4.3	1.2	1.5
Mo	0.5	0.8	1	1.5	1.5	1.4	0.8	1
Zn	90	110	75	50	33	30	80	60
Pb	12	21	26	30	36	40	40	20
Zr	55	70	130	170	200	150	800	200
Hf	0.2	0.6	1	1.5	5	7	21	1
Nb	3	4	7	17	23	80	120	20
Ta	0.3	0.6	0.8	1.5	3.6	10	6	3.5
REE+Y	40	60	120	180	260	350	500	185
U	0.8	1.5	2	4.6	8	9	10	3.5
Th	4	5.3	7.5	25	38	40	42	18

N.B. The table is based on the average contents of elements as derived in the work of S.M. Beskin et al. (1979) [4].

Table 2.2 Characteristic Mineral Associations and Geological Setting of Amazonite in Granitoid Formations

		Formation type					
		Alaskite		Subalkaline-leucogranite		Alkaline-granite	
Formation depths of plutons (km)	Age of granitoid formations	Primary and secondary (in parentheses) genetic types of rocks[a], region, district	Characteristic association of minerals	Genetic types of rocks[a], region, district	Characteristic association of minerals	Genetic types of rocks[a], region, district	Characteristic association of minerals
4–7	Proterozoic to early Paleozoic	5[a]; Canada, Québec, Ville-nieve	Muscovite, garnet, schorl, apatite, fluorite, uraninite, monacite	3; Canada, Québec, Leduc	Lepidolite, tourmaline (most frequently polychromatic), columbite-tantalite	5; Russia, Sliudanskoe, eastern Siberia	Biotite, fluorite, fergusonite, euxenite, betafite, titanite
		5; India	Muscovite, biotite, beryl, garnet, schorl, apatite, uraninite, fluorite	3; Sweden: Utö	Lepidolite, tourmaline (most frequently polychromatic), columbite-tantalite, spodumene, ambligonite, petalite	5; Russia: Kola Peninsula	Yttro-fluorite, gadolinite, yttrialite, columbite-tantalite, pyrochlore-microlite, galena
		5; Russia: western Baikal region	Biotite, muscovite, garnet, schorl, apatite, rutil	3; USA: North Carolina	Lithium mica, tourmaline (frequently polychromatic), spodumene, uraninite	5; Russia: northern Baikal region	Fluorite, hadolinite, columbite, fergusonite, samarskite, priorite
		4; Russia: Azov region	Biotite, topaz, fluorite	3; Madagascar	Lepidolite, tourmaline (frequently polychromatic), beryl, spodumene	5; Russia: lakutiia	Zircon, yttrialite, fergusonite, pyrochlore, galena
2–4	Caledonian	4; Kazakhstan: Zerendin	Fluorite, morion, smoky quartz, apatite, rutile	2, 4; Kyrgyzstan	Zinnwaldite, topaz, fluorite, columbite-tantalite	6; Ukraine	Fluorite, cytolite, columbite, bastnäsite
		6; Kazakhstan: Zolotonosh	Schorl, topaz, columbite, cassiterite, rutile	2, 4; Egypt, Algeria	Lithium mica, topaz, columbite-tantalite, cassiterite, thorite, fluorite	6; Kazakhstan: Losevskoe	Fluorite, biotite, cyrtolite, columbite, ilmenite

Era			
2–4 Hercynian	Biotite, muscovite, columbite, samarskite, fluorite 4, 6; Kazakhstan: Kyngyrzhal'	Protolithionite, zinnwaldite, topaz, columbite-tantalite, cassiterite, fluorite 2, 4, (5, 6); Kazakhstan: Maikul'	Topaz, phenakite, beryl, columbite-tantalite, fergusonite, fluorite, cryolite 5; Russia: Urals, Il'menskii Mountains
	Biotite, fluorite, schorl, apatite, rutile 4, 5; Tajikistan	Zinnwaldite, topaz, columbite-tantalite, cassiterite 2, 4; Kyrgyzstan: Khorgos	Topaz, beryl, columbite, fergusonite, betaphite, samarskite, fluorite 5; Madagascar
	Biotite, lithium muscovite, beryl, schorl, fergusonite, orthite, fluorite 4; Poland: the Sudetes	Protolithionite, zinnwaldite, topaz, columbite-tantalite, cassiterite 2, 4; Russia: eastern Saiian	Topaz, beryl, phenakite, helvine, fergusonite, columbite, pyrochlore, cryolite 5; USA: Pikes peak, Amelia
1–3 Cimmerian	Biotite, topaz, muscovite, schorl, fluorite 5 (4); Russia: Transbaikal, Adun-Cholon	Protolithionite, zinnwaldite, topaz, columbite-tantalite, pyrochlore-microlite, cassiterite, fluorite, galena 2, 4, 5, 6; eastern Transbaikal	Biotite, topaz, tourmaline, columbite, zircon, fluorite, ilmenite 5 (4); Mongolia: Dzun-Bani
	–	Protolithionite, zinnwaldite, topaz, columbite-tantalite, pyrochlore-microlite, cassiterite, fluorite 5; Mongolia	–
< 1 Alpine	–	Lepidolite, tourmaline (polychromatic), topaz, columbite-tantalite, hambergite 5; Tajikistan: Pamirs	–

aGenetic types of rocks: (1) subvolcanic vein granitoids (primarily ongonites), (2) amazonitic granites (and aplites), (3) pegmatite veins not found to have connections with granitoid plutons, (4) pegmatite bodies and similar formations in endocontacts of plutons, (5) pegmatite bodies in near and distant exocontacts of plutons, (6) local (endo- and exocontact) metasomatites and (amazonite) feldspar-quartz veins.

In the genetic types of leucogranite-alaskite formations, amazonite is relatively rare and is noted in rare-metal-muscovite pegmatites of medium to large depths, which were formed in the late Proterozoic to early Paleozoic and do not exhibit a close association with concrete intrusions (India, Canada). Amazonite is also noted in pegmatites from endo- or exocontacts of alaskite plutons with rare-metal (non-lithium) and/or topaz-fluorite miarolitic mineralization, which were formed at middle and small depths in the interval from the Precambrian to the Mesozoic (Ukraine; Adun-Cholon, Kazakhstan). Alaskite plutons of this subtype can be accompanied spatially by alkaline granites, in the endocontact metasomatites[1] of which, likewise, is noted amazonite (for example, Losevskii in the boundaries of the heterogenic Zerendinskii pluton). It is known also in the endocontact metasomatites and feldspar-quartz veins of albitite-greisen-bearing plutons (Zolotonosh and Kyngyrzhal' plutons in Kazakhstan) with columbite, cassiterite, and other minerals. In plutons with typical quartz-vein-greisenic mineralization, occurrences of amazonite have not been recorded.

Of all the amazonite-containing genetic types of the alaskite formation, the weakest amazonitization is exhibited by certain miarolitic pegmatites of endocontact zones of alaskitic plutons, while fairly intensive amazonitization is occasionally common to nonintrusive pegmatites with rare-metal-muscovite mineralization (Indian type).

The subalkaline-leucogranite formation is the undisputed leader among amazonite-containing granitic formations. Here are developed all possible genetic types of amazonitic rocks. To the earliest—from Precambrian to late Paleozoic—and to the deepest formations are associated rare-metallic sodium-lithium pegmatites (Leduc, Canada; Utö, Sweden; Bastar, India; and elsewhere) with the characteristic association of minerals: lepidolite, polychromatic tourmaline, and beryl, and sometimes spodumene and others.

Amazonitic granites and their other genetic types were formed at a relatively late time (from the Paleozoic to the Mesozoic inclusively) and at comparatively small depths; moreover, the largest plutons of such granites usually are also the deepest, and the establishment of small bodies accompanied by ongonites (Mongolia) evidently occurred at minimal depths (less than 1 km) at the most recent time.

By degree of incidence, amazonite of the granite of the examined formation is exceeded many times over by all other genetic types (the sizes of certain plutons reach 15 km^2), but the intensity of amazonitic color of granites is lower than

---

[1]It should be emphasized that cases of similar telescoping enable a fairly confident prediction for the development of amazonitization, produced by the genetic types of alkaline-granite formation, in the more ancient pegmatites of the alaskite formation (which includes also moderately mica-bearing pegmatites); moreover, the interruption in time between the establishment of pegmatites and the superposition of amazonitization can be fairly great: up to a few hundred million years. Examples of such polyformational pegmatites might be found in certain amazonite-containing bodies of the Kola Peninsula, Karelia, Vyborg district, and the Baikal region in Russia, and possibly in India.

that of pegmatites and, as in others, consistently increases from internal, weakly amazonitized zones of granites to domed zones, and consequently, the color is more intensely rendered in feldspars from pegmatoid zones of endocontacts (stockscheiders).

Pegmatites of sodium-lithium type, for which a direct connection with granites has not been established, are characterized by a highly weak green color of amazonite; sometimes researchers either do not notice it at all, or they report green or greenish microclines without identifying them as amazonites. On the basis of a limited number of examples of pegmatites of a similar type, some strengthening of the color of amazonites may be proposed in the most ancient and deepest belts. However, more frequently in pegmatites of the said type, especially those containing spodumene, ambligonite, or pollucite, amazonite generally does not occur.

Amazonite is fairly typical also for the alkaline-granite formation, but here its development is predominantly to be found in pegmatites of which the lower age boundary is dated most probably from the Proterozoic to the early Paleozoic, and the upper boundary to the beginning of the Mesozoic. Corresponding with the age and proposed depths of formation, and also with the structural-textural particularities, mineral composition, and other data is noted a definite change of types of pegmatites of this formation: from rare-metal/rare-earth (Kola Peninsula, Baikal region, Iakutiia, eastern Siberia) to the primarily rare-metal (Il'men Mountains, Russia; Pikes Peak, USA; and elsewhere). Characteristic particularities of pegmatites of the latter type are the presence of cavities with colored stones and wide distribution of graphic textures. In pegmatites of the examined formation, amazonitization is developed nonuniformly and to a minimal degree in pegmatites of close exocontacts of zones of alkaline granites; the greatest development of amazonitization is reached in separate types of pegmatites of distant exocontact zones of plutons at a certain optimal distance from the latter (in concrete instances, these intervals range from several hundreds of meters to one or sometimes several kilometers); at a great distance from granites, amazonitization again gradually disappears.

Significantly more rarely, in a small quantity and with relatively pale color, amazonite occurs in metasomatites of endo- and (more frequently) exocontacts of plutons of alkaline granites (Losevskoe in Zerendinskii pluton, Perga cluster in Ukraine, and others). Analogous exocontact changes may also accompany separate bodies of alkaline-granite pegmatites (Kola Peninsula, Russia; Prilepy, Montenegro).

Summarizing the information on the prevalence and scale of development of amazonite in rocks of various formations, it may be noted that it is most common for the subalkaline-leucogranite formation, where it is noted in practically all genetic types, although its most significant occurrences should be considered amazonitic granites in particular. This formation markedly exceeds the alkaline-granite formation, in which amazonite is most common for pegmatites. Finally, in rocks of the alaskite formation, amazonite is fairly exotic and, moreover, connected with the most alkaline facies of its granites. In terms of the number of finds and of the intensity of color of amazonite, pegmatites play a basic role among representatives of the alaskite formation, but in terms of its degree of incidence, they are inferior to endocontact metasomatites.

The age of rocks in which amazonite occurs varies within a highly wide range: from Proterozoic to Alpine. To the most ancient (Proterozoic) are dated the amazonitic pegmatites of the alkaline-granite formation (Kola Peninsula, eastern Siberia), in which amazonitization sometimes reaches a maximum of intensity as well as of uniformity of manifestation of color; and to the youngest (Jurassic to Cretaceous) are dated certain granites and their other genetic types of the subalkaline-leucogranite formation (Transbaikal, Russia; Mongolia), where amazonitization encompasses entire plutons, but the intensity of color of this mineral (even of amazonitic potassium feldspars from pegmatoid and quartz veins, directly connected with granites) always is somewhat weaker than in amazonitized bodies of pegmatites of the alkaline-granite formation of the Kola Peninsula.

Evaluating the temporal distribution of amazonite, it is easy to notice a significant increase toward the younger epochs as much in degree as in its scale of occurrence. On the other hand, the most intensive amazonitization, evaluated in terms of intensity, variability of specific color, and regularity of development, accrues to the most ancient pegmatitic veins. Of interest is also the differentiated examination of the temporal distribution of amazonite in terms of each of the granitoid formation types.

For the alaskite type, from ancient to young epochs, is noticed an insignificant increase in the number of amazonite-containing objects and more significant increase in the scale of occurrence of amazonitization (principally at the expense of endocontact metasomatites), accompanied by some decrease of intensity of amazonite color.

In the formations of the subalkaline-leucogranite type, toward the younger epochs, there is a marked increase in the degree of distribution as much as in the scale of amazonitization. The tendency of the change in intensity of the color is unclear (a certain weakening of intensity is noted for sodium-lithium pegmatites and pegmatites from endocontact zones of plutons, and a strengthening for amazonitic granites).

For alkaline-granite types, the directionality of the change in distribution (apparently also an increase) has not been identified, but the intensity of coloration of the younger amazonites clearly decreases in pegmatites as in other bodies of this formation.

The focus turns now to the character of the change of amazonitization for near-coeval and single-depth genetic types of rocks from various granitoid formations. For amazonite-containing pegmatites of the Proterozoic (early Paleozoic) that were formed at large to medium depths, the maximum of intensity of amazonitic coloration accrues to the alkaline-granite formation; the same may be confirmed also for Caledonian pegmatites with amazonite (medium to small depths). In Hercynian and even more for the Cimmerian epoch, the maximum of amazonitization shifts in the subalkaline-leucogranite formation, where amazonite can occur in all genetic types. Thus, contrasted with the common regular growth of scale of amazonitization in time, accompanied with some dilution of color, the role of the initially primary concentration of amazonite is lost by pegmatites of the alkaline-granite formation, and altogether more significance is acquired by a wide range of genetic types with which amazonite is associated in the subalkaline-leucogranite formation. In other words, if in ancient epochs the optimal conditions for the manifestation of amazonite arise only in pegmatites of the alkaline-granite formation, then in the younger epochs, such conditions are assured in rocks of various genetic types of all three formations, although the maximum opportunity in this regard is unarguably characterized by genetic types of the subalkaline-leucogranite formation.

In summary, it should be emphasized once more that amazonite occurs predominantly in granites, pegmatites, and metasomatites that are related to genetic types of alaskite, subalkaline-leucogranite, and alkaline-granite formations. Some essential particularities of these formations will be examined below.

2.3 MAIN GENETIC TYPES OF AMAZONITE-CONTAINING ROCKS

2.3.1 Amazonitic Granites and Their Other Genetic Types

Amazonitic granites, initially regarded as an exotic type of granitoids, are treated in the current literature as fairly widespread geological formations important in many aspects. According to A.A. Beus and V.I. Kovalenko et al., the principal value of amazonitic granites is derived from the fact that, in a range of cases, they are industrial sources of tantalum, niobium, and tin, and a prospective source for obtaining lithium, rubidium, cesium, and other elements. However, not all geologists subscribe to this opinion on their potential ore-bearing and search significance [2,47]. This reluctance is in all likelihood connected with the fact that to these researchers, the presence of several subtypes of amazonitic granites with fairly sharply differentiated degree (and character) of ore-bearing remained unknown and consequently fell outside the focus of their attention [4].

By the present time, amazonitic granites have been described in great detail in the works of A.A. Beus, L.G. Fel'dman, S.M. Beskin, V.I. Kovalenko, and Ia.A. Kosals et al. Therefore, the author et al. [47] have set it as their task to summarize all obtained information and to succinctly characterize that genetic type of rare-metal deposits.

Rare-metal deposits of this type now are widely known in Russia (Transbaikal), in CIS countries (Kazakhstan and Kyrgyzstan), and in a range of other countries (Algeria, Egypt, China, Mongolia, and USA). According to the data of A.A. Beus et al., plutons of similar granites are grouped in linear zones by length from 3 km to several dozen kilometers, by width from 0.5 to 10 km. These zones usually coincide with belts of dikes and of minor intrusions of various composition and age, and to regional faults. The correlation of linear zones of distribution of subalkaline granites with geological structures is frequently cross-cutting, rarely conformable. Subalkaline granites constitute dike and stock bodies (steep or inclined occurrences) and flat-lying deposits, irregular-isometric in plan. The length of dike and stock plutons in plan varies from between 10 and 100 m up to several hundred meters or a few kilometers, and the width from 5 m to several hundred meters. In depth, they have been tracked from several dozen to several hundred meters. Flat-lying plutons have a size in plan from 0.5 by 0.5 km or 1 by 3 km with vertical thickness from between 1 and 10 m up to 80 m.

Usually present in the development area of subalkaline granites are manifestations of earlier granitoid formations (granite and alaskite-leucogranite). Alaskites, as a rule, represent plutons of one of the subtypes: either common greisen-bearing or rare-metal-pegmatite-bearing.

The enclosing rocks of plutons of the subalkaline-leucogranite formation are igneous-sedimentary rocks as well as granitoids of earlier formations. The relationship of plutons of amazonitic granites to structures of enclosing rocks is sharply discordant in the case of steep occurrences, but can be conformable in the transition to flat-lying forms. Furthermore, intrusions not infrequently cleave along the upper contact surface of the section between the earlier granitoids and their enclosing rocks. Alongside intrusions of amazonitic granites are usually observed more intensive contact-metasomatite variations of enclosing rocks. Thus, in the more ancient granitoids arise zones of biotitization, metacrystals of quartz, albitization, and amazonitization, which not infrequently create an illusion of gradual transitions between the granitoids of earlier formations. The sizes of plutons composed entirely of amazonitic granites as a rule are not large (1–1.5 km²). The age of granitoid plutons with amazonite is defined from middle Paleozoic (rarely early Paleozoic) to the late Mesozoic. In all cases, amazonitic granites are accompanied by a series of their other genetic types: aplite- and granite-pegmatites, pegmatoids, and quartz veins with amazonite.

The characteristic and general particularities of amazonitic and amazonite-containing granites are consistently reported as increasing in degree of amazonitization from the internal weakly amazonitized zones of granite intrusions toward the domed zones. Consequently, the most intensively colored are the potassium feldspars in vein derivatives and in metasomatites that are embedded in endo- and exocontacts of granites.

Amazonite-containing granites are characterized, as a rule, by zoned texture. Most typical is the following type of zonality (from the center of a pluton to the margins): biotite-muscovite—microcline-albite—amazonite-albite—lithium mica. Occasionally, it can be fairly difficult to differentiate zones of microcline-albite and amazonite-albite granites; in such cases, the transitions of biotite granites to amazonite (microcline)-albite are, likewise, virtually gradual.

More rarely have been noted plutons composed entirely of amazonite-albite granites with lithium mica, which are particularly widely developed in peripheric

toward their center of two to three varieties by indistinctly exposed, but distinct textures (graphic, apographic, and block). The main rock-forming minerals of pegmatites are microcline, pale-green amazonite, oligoclase, albite, and quartz.

Very rarely, insignificant amazonitization of potassium feldspar that encloses cavities with crystal fluorite and pure quartz is reported in miarolitic pegmatites located in crystal-pegmatite-bearing intrusions (Kazakhstan). Endocontact pegmatites were formed primarily in the Paleozoic and are distinguished by a dependence on formation depths of muscovite-topaz or biotite-fluorite (Ural) mineralization [47].

Amazonitic pegmatites of the subalkaline-leucogranite formation are not associated directly with granitoid plutons and are embedded in all manner of metamorphic rocks of Precambrian and more rarely of Paleozoic age: granitogneisses, schists, amphibolites, quartzites, and others. Pegmatite bodies have dike-vein form, are of medium size (length, from a few hundred meters to several kilometers, thickness of several meters, in rare cases up to 10 m), and are weakly differentiated.

The mineral complex associated with greenish-yellow and greenish amazonitic potassium feldspar is represented as lepidolite, polychromatic tourmaline, and beryl, and rare tantalum-niobium oxides (Canada, Sweden, USA). In rare cases and in insignificant quantities, spodumene is noted here, but, as mentioned already, amazonite is not typical for microcline-albite and albite pegmatites with lithium minerals (of sodium-lithium type).

Most interesting and important in practical terms are amazonitic pegmatites of the alkaline-granite formation. Characteristic for them are a wide and intensive development not only of amazonitization, but also of other post-magmatic processes (albitization, greisenization, and ore formation). According to the data of I.V. Bel'kov and A.Ia. Lunts, pegmatite bodies with amazonite are localized predominantly in the far and near exocontacts of alkaline plutons and are confined to partially transformed alkaline dissolution or unaltered gneisses, schists, and other metamorphic rocks.

The most ancient amazonitic pegmatites of Precambrian age with rare-earth mineral paragenesis are known at present in many regions of the globe: in Russia on the Kola Peninsula and in eastern Siberia including the Baikal region, the USA, Nigeria, Mozambique, and elsewhere [7,47]. The form of these pegmatite bodies is most frequently lentiform-vein, and their length ranges from a few dozen to a few hundred meters, more rarely up to 400 to 600 m, with thickness from a few meters up to 15 to 20 m in the bulge of veins. Zonality in pegmatite bodies is expressed fairly weakly and is reported only in terms of the increase in size of individuals of rock-forming minerals in the direction from the selvage toward the center of veins and in the change in the same direction of the color tone of potassium feldspar from gray and pink to bluish-green of various hues and saturation. Apart from microcline and amazonite, the principal minerals of pegmatites include albite and quartz, the secondary minerals include biotite, fluorite, magnetite, garnet, and the accessory minerals include gadolinite, tantalite-columbite, genthelvite, fergusonite, cyrtolite, and galena, and more rarely yttrialite, thorite, pyrochlore, samarskite, and beryl.

The early to mid-Paleozoic amazonitic pegmatites are fairly close to those described above in terms of form, sizes, and internal structure, but they are distinct in terms of their accessory mineralization: gadolinitic with beryl (Iakutiia, Russia; southern Norway) or only beryllic (western Baikal region and eastern Saiian, Russia; Poland).

Amazonitic pegmatites of late Paleozoic age associated with the alkaline-granite formation are known in Russia in the Ural and in Tuva, as well as in Madagascar and in the USA. Characteristic for these are platy and lentiform forms of bodies, short length (as a rule, of several dozen meters) and thickness (up to 5 or 10 m). Pegmatite veins are differentiated, and their larger part is occupied by zones of graphic and pegmatoid texture, which sometimes contain large cavities with crystals of feldspars, quartz, topaz, and several other minerals. Rock-forming minerals of these pegmatites are represented by microcline, amazonite, albite, and quartz; the secondary minerals are micas; and the accessory minerals are garnet, topaz, beryl, phenakite, columbite, fergusonite, samarskite, helvite, and thorite.

2.4 MAIN PROVINCES AND DEPOSITS OF AMAZONITE

2.4.1 Kola Peninsula

At the end of the 1920s, in the central part of the Kola Peninsula were discovered very large plutons of alkaline granites and their associated amazonitic pegmatites. Further research of this district has enabled the identification of an entire range of pegmatite fields situated in the exocontact zones of plutons of alkaline granites.

Among the fields of development of amazonitic pegmatites, the largest and most typical in terms of geological structure is the western Keivy. Its geological structure is constituted by various magmatic, metasomatic, and metamorphic rocks, among which the greatest distribution, according to the data of many researchers, accrues to the Proterozoic and the smallest to the Archean and Paleozoic formations (Fig. 2.2).

To the Proterozoic rocks belongs a thick series of sedimentary–metamorphic rocks of early Proterozoic age (the gneissic and schists complex of the Keivy series), intrusions of meta-gabbro, late-Proterozoic large intrusions of alkaline granites and their metasomatites, and nepheline syenites, as well as various vein rocks (pegmatites, aplites, and quartz veins).

The alkaline rocks most widely developed within the boundaries of the pegmatite fields—granites, syenites, and miaskites—belong to the late Proterozoic. The western Keivy pluton of alkaline granites forms a large intrusion of zoned structure. It is formed in the center by medium- to coarse-grained aegirine-arfvedsonite and arfvedsonite granites, and in the intermediary and peripheral

N

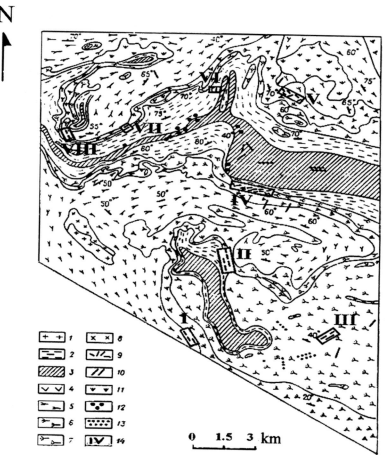

FIGURE 2.2

Schematic geological map of the district of the western Keivy. According to V.V. Nosikov et al: (1) plagio-microclinic granites and gneissogranites of the Archean; (2–8) Proterozoic (2: rocks of gneissic complex, 3: rocks of schists series, 4: amphibolites, 5: alkaline granites, 6: alkaline-granite gneisses, 7: alkaline gneiss-metasomatites, 8: syenites and miaskites); (9) rare-earth and amazonite-containing pegmatites (depicted in frames); (10) dikes of diabase, gabbro-diabase, and gabbro-norite; (11) Quaternary deposits; (12) beryl-muscovite pegmatites; (13) quartz veins, (14) sections with zoned layout of various pegmatites.

zones is formed by the medium-grained aegirine and arfvedsonite as well as fine-grained aegerine granites.

The processes of intensive alkaline metasomatism that accompany the injection of the intrusion have significantly altered the enclosing metamorphic rocks, which are transformed into metasomatites varying in terms of composition.

To the western Keivy alkaline granites spatially coincides a small pluton of alkaline and nepheline syenites, tending toward the contacts between oligoclase granitogneisses and alkaline granites.

Pegmatites. The western Keivy pegmatite field consists of 15 sections, within the boundaries of which are observed pegmatitic veins that vary in terms of composition and geological position. These sections spatially coincide with the late-Proterozoic pluton of alkaline granites. Pegmatite veins are widely developed inside the alkaline intrusion, and likewise in its endo- and exocontact zones. In the coarse- to medium-grained arfvedsonite and aegirine-arfvedsonite granites, as well as in the fine-grained endocontact aegirine granites and gneissogranites are embedded microclinic pegmatites with aegirine and arfvedsonite.

Pegmatite bodies have lentiform and vein form, small sizes, and are weakly differentiated. The outer zones of vein bodies are composed of quartz-microclinic pegmatite of granitic texture with arfvedsonite, and in the central part by pegmatite of segregational texture. Occasionally, pegmatites, localized in fine-grained gneiss-granites, contain pale-colored (partially amazonitized) microcline, which in pre-selvage sections gives way to pink potassium feldspar. A characteristic feature of these pegmatites, according to the data of A.Ia. Luntz and M.N. Ostrooumov, is the total absence of rare-earth (REE) mineralization.

Coinciding with the exocontacts of alkaline pluton and with the rocks of zones of alkaline metasomatism are the main deposits of pegmatite veins of the district. In form, these bodies are nearly indistinguishable from the microclinic pegmatites scattered among the alkaline granites. They are traced along the strike at a distance from several dozen meters up to 800 or 1000 m with thickness from 0.5 or 5 m up to 10 or 20 m in bulge of the bodies. In the peripheral zones of the vein bodies is observed plagioclase (albite-oligoclase)-quartz or quartz-microclinic pegmatite of granitic, more rarely graphic, texture, which in the inner zones is replaced by quartz-albite-microclinic pegmatite of pegmatoid and block texture. In rare cases, in the central parts of veins may be seen the axial zone of monomineral quartz. Equally rarely, potassium feldspars of these pegmatites are weakly amazonitized. The specifics of pegmatites confined to the near exocontact of alkaline pluton consist of the widespread development in them of processes of greisenization and albitization, as well as the presence of notable concentrations of rare-earth minerals.

Amazonitic pegmatites are situated at a great distance from alkaline granites.

Furthest from the granite pluton are situated microcline-plagioclase and plagioclase-microcline pegmatites with muscovite and accessory beryl, which are embedded in mica schists of the Keivy series. Particular to pegmatite veins of this type are an extremely weak intensity of processes of albitization and amazonitization and the practically total absence of rare-earth minerals.

Amazonitic Pegmatites. Typical amazonitic pegmatites are localized in injection and unaltered biotite and garnet-biotite gneisses, more rarely in gneiss-metasomatites. In terms of the character of occurrences and of the form of vein bodies, amazonitic pegmatites are indistinguishable from microclinic pegmatites. Sizes of pegmatite veins with amazonite range from a few dozen to a few hundred meters with a thickness from several meters to 15 or 20 m in the bulge of veins. The contacts of pegmatites with enclosing rocks are sharply straight, rarely wavy. Near-contact alterations are expressed weakly, but sometimes exo-contact sections are intensively biotitized, albitized, and amazonitized. The pegmatites are undifferentiated or weakly zoned, as expressed in the gradual coarsening of granularity and an insignificant change in mineral composition in the direction from the periphery to the center of veins. The transitions between textural zones are indistinct and gradual. The marginal zones of pegmatite bodies are composed predominantly by quartz-albite-oligoclase and quartz-albite pegmatite of granitic texture, which in intermediary zones are replaced by quartz-microcline (rarely quartz-amazonitic) pegmatite of graphic texture or microclinic (amazonitic) pegmatite of pegmatoid and block textures (Fig. 2.3). Occasionally, observed in the center of veins is a discontinuous zone of quartz nucleus. According to the quantitative correlation of microcline and amazonite among amazonitic pegmatites can be distinguished amazonite-microclinic and microcline-amazonitic types.

The pegmatites of several sections of the western Keivy pegmatite field have been explored in detail. Within the boundaries of one of them (Fig. 2.4) have been encountered representatives of all groups of pegmatites described above. Judging by the available materials, there is a particularly prominent occurrence here of a zoned distribution of mineralized pegmatite veins in the exocontact zone of the alkaline-granite intrusion. Thus, it was established that the quartz and weakly albitized microclinic pegmatites with accessory rare-earth mineralization (veins 16 and 42) are situated in the near exocontact of the pluton (no closer than 50 m and no further than several hundred meters), whereas intensively albitized microclinic pegmatites with a large concentration of rare-earth

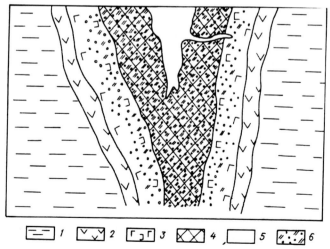

FIGURE 2.3
Schematic transverse cross-section of pegmatite body of amazonite-microclinic composition: (1) biotitic gneisses; (2–4) pegmatite (2: plagioclase-quartzic of granite texture, 3: quartz-microclinic of graphic and apographic textures, 4: microclinic (amazonitic) of block and pegmatoid textures); (5) zone of monomineral quartz; (6) sections of intensive albitization and greisenization.

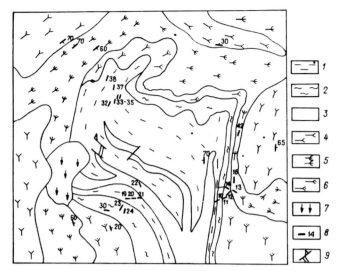

FIGURE 2.4
Schematic geological structure of pegmatite sections. According to A.P. Gavrilov et al: (1–2) gneisses (1: biotitic and garnet-biotitic, 2: alkaline injection); (3) mica schists; (4–6) alkaline rocks (4: granites, 5: granitogneisses, 6: gneisso-metasomatites); (7) Quaternary deposits; (8) rare-earth and amazonitic pegmatites, numbers of veins; (9) zones of tectonic dislocation.

minerals are embedded at 500 to 800 m from the granite pluton (veins 11–14; see Fig. 2.4). Albitized amazonite-microclinic pegmatites with a small quantity of accessory rare-earth minerals (vein 10) are located at a distance of approximately 1 km from the granite, while insignificantly albitized and weakly mineralized microcline-amazonitic pegmatite veins are separated from the pluton at 2 to 3 km (veins 19–21, 23, 33–35, and others). Still further from the alkaline granites (up to 5 to 8 km) are localized oreless (practically without amazonite) plagioclase-microclinic pegmatites with muscovite (vein 22).

Amazonitic pegmatites of this section tend toward biotite, in places amphibole-biotite gneisses of the Lebyazhinsk suite of the Keivy series and by composition are associated with the microcline-amazonitic type. In one of the pegmatitic veins, all potassium feldspar is represented by amazonite (vein 19). This pegmatite body has regular lentoid form (traced along the strike up to 200 m with an established thickness from 10 to 20 m in bulge). The marginal zones of the vein are composed of fine-grained plagioclase (albite-oligoclase) pegmatite of granite texture with quartz, with small grains of nonuniformly grained amazonite (Fig. 2.5). This further follows the zone of principally amazonitic pegmatite of pegmatoid and block textures, which in the center gives way to a zone of quartz composition of nonrecurrent thickness. Evidently, it is no accident that pegmatite of graphic and apographic textures remain practically unobserved in this pegmatite body, while pegmatite zones of block texture predominate. In the opinion of the author et al. [47], the highly intensively developed recrystallization of potassium feldspar serves as one of the favorable conditions for the unique scale of amazonitization encompassing even the enclosing rocks (Photo 1).

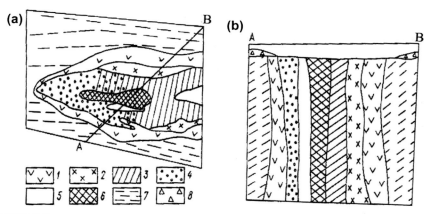

FIGURE 2.5
Schematic plan (a) and cross-section (b) of the western flank of vein 19. According to M.N. Ostrooumov.
(1–2) pegmatite (1: plagioclase-quartzic of granite texture, 2: albite-amazonitic of pegmatoid texture);
(3–5) amazonitic (3: light-blue of pegmatoid and fine-block textures, 4: green of fine-block texture, 5: dark-green of block texture); (6) zone of monomineral quartz; (7) biotitic gneisses; (8) quaternary deposits.

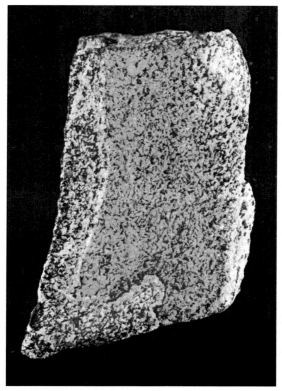

PHOTO 1
Amazonitization of the gneisses enclosing a pegmatite vein (Ploskaia Mountain, western Keivy pegmatite field).

Rock-forming minerals of pegmatitic veins are represented by amazonite, albite, and quartz, and the secondary minerals by mica (biotite, sometimes zinnwaldite). Among the accessory minerals, according to the data of a range of researchers, the most common are tantalum-niobates oxides (columbite-tantalite, Pb-pyrochlore-microlite), fergusonite, gadolinite, cassiterite, cyrtolite, fluorite, galena, and native bismuth and its sulfides and sulfosalts.

Variously colored amazonites are zonally distributed in this pegmatitic body, representing an industrial deposit of amazonitic feldspar: toward the axial and lower parts of the vein are found predominantly amazonites with green hues of colors, and in the pre-selvage and outer zones as well as in the upper parts of the vein are observed primarily potassium feldspars with blue amazonitic colors. In association with the indicated distribution of variously colored amazonites also occurs a change in the quantity, sizes, and form of perthitic ingrowths of albite. Furthermore, from the observation of direct intersections of multicolored amazonites, it can be concluded that the green amazonites on the whole were formed later than the blue ones. We note also that in similar pegmatites, where the process of amazonitization achieves the maximum (these veins are devoid even of relics of common potassium feldspars), other post-magmatic processes (greisenization, albitization, and ore mineralization) have a much lesser intensity in comparison with the pegmatite bodies in which amazonite is less widely developed.

But nevertheless, in this as well as in other sections of the western Keivy field, the greatest distribution is found among pegmatites of amazonite-microclinic composition, in which common microcline in the majority of cases predominates over amazonite. Amazonite in these pegmatites is localized primarily in the central parts of the vein bodies, and microclines of common colors (white, gray, pink, etc.) are most frequent in peripheral and intermediary zones. As a rule, between microcline and amazonite are observed entirely gradual transitions—it is clearly visible as the yellowish-pink potassium feldspar initially acquires a barely notable blue hue, which later toward the inner zones of pegmatite bodies incrementally becomes all the more saturated and more intensive. It is difficult to establish precisely the border of this transition, and for this reason, the impression is created that the amazonitization of a microcline bears a variable character, with vague contours of stains and isolations.

The highest intensity of amazonite color is particular to potassium feldspars from axial zones of pegmatite bodies, where blocks of amazonite (up to 1 to 2 m) are formed by microclines fairly uniformly colored in bluish-green and green of hue. In the peripheral and intermediary zones of the granitic, graphic, and block textures, amazonite does occur (more widely developed here is common microcline), although its color sometimes does not reach the brightness and saturation characteristic for amazonites from pegmatoid and block textures found in the inner parts of the pegmatite bodies.

Observations of the distribution of variously colored potassium feldspars and amazonites within the boundaries of the uncovered part of the industrial pegmatite body and neighboring nonindustrial veins, situated on a higher hypsometrical level, allow the hypothesis that the intensity of development of amazonite gradual decreases along the transition to the upper and peripheral sections of the vein bodies. Along with amazonite, potassium feldspars with common colors occur (in various quantities) in practically all the explored pegmatite veins (the content of the first ranges from 0% to 70%, of microcline from 5% to 90%).

The presence of notable concentrations of rare-earth and rare-metal minerals on the whole is not particular to these pegmatites; nevertheless, it can be noted that the accessory yttrium-ytterbium and tantalum-niobium mineralization fairly often is concentrated precisely within the amazonite-containing pegmatites. Attracting attention is the presence in them of bismuth (partially tin) mineralization that is uncommon for pegmatites. As the result of detailed mineralogical studies conducted in 1986 by A.V. Voloshin et al. [51], uncovered in these pegmatites was an entire series of new accessory rare-earth minerals (keivite, vyuntzpakhkite, caysichite, and others) and tantalo-niobates (tantite and others).

Summarizing the above, pegmatites distinguished by the quantity of amazonite in vein bodies and the degree of microcline amazonitization process are zonally distributed throughout the alkaline-granite pluton. The maximally mineralized microcline pegmatites (sometimes with pale-colored amazonite, but more frequently without it) are located in direct proximity to the granites, while microcline-amazonitic pegmatites with accessory mineralization are situated at a significant distance from the granites. Between them, the intermediary position is occupied by mineralized amazonite-microcline pegmatites. Rare-earth and rare-metal mineralization is practically absent as much in the microclinic pegmatites embedded directly in the alkaline granites as is in the microcline-plagioclase pegmatites localized at a maximum distance from the alkaline rocks and the tending toward schists Keivy series.

Thus, geological observations demonstrate that the maximum occurrence of the amazonitization process (expressed in the development of microcline-amazonite, and in isolated cases also amazonite pegmatites) is evidence of the existence of insignificant concentrations of rare-earth and rare-metal minerals directly within these pegmatite bodies. It is evident that pegmatites of this type serve primarily as a principal source of extracting amazonite. Additionally, finds of microcline-amazonite pegmatites indicate the possibility of discovering ore-bearing pegmatite veins situated in lesser, not primarily amazonite pegmatites, at a distance from their generating alkaline plutons. Also of interest is the indicatory role of amazonite in relation to muscovite-beryl (with accessory beryl) mineralization in the pegmatites most distant from the pluton. Thus, the prospecting significance of amazonite becomes understood; the type and metallogenic specialization of vein bodies associated with alkaline-granitic plutons may be evaluated according to amazonite's quantity and quality in pegmatites.

2.4.2   Il'menskie Mountains

The district of the Ilmen reserve (southern Ural), composed of various metamorphic and igneous rocks of Paleozoic age, has been explored in detail. At the base of the cross-section of metamorphic rocks are embedded paragneisses with interbedding of amphibolites, as well as paragneisses, quartzites, schists, and amphibolites, joined correspondingly in the Selyanka assemblage of rocks and in the Vishnevogor suite. Rocks along the cross-section of the Ilmen suite are developed in the wings of the Ilmen anticline, in the nucleus of which are observed rocks of the Selyanka assemblage. By composition, the Ilmen suite is essentially amphibolitic. Uppermost along the cross-section, the Argayash and Ishkul' assemblages were formed by schists, quartzites, and gneisses varying in terms of composition.

Igneous rocks of this district are represented predominantly by miaskites and granitoids, and to an insignificant degree by ultrabasite and gabbroids. Miaskites and their associated alkaline rocks constitute the large Ilmenogorskii pluton in the southern part of the Il'menskie Mountains and the three narrow submeridional bands in the north. The composition of the Ilmenogorskii pluton includes biotite, biotite-amphibolite, and amphibolite miaskites, syenites, and phenites. The age of the miaskites is 310 ± 10 Ma.

Within the Ilmenogorskii complex are distinguished two stages of granitization: early plagiogranitic (approximately 435 ± 44 Ma) and late granitic (261 ± 28 Ma). The products of the second stage are distinctly post-miaskitic and are represented primarily by small bodies of granites, as well as aplite and pegmatite veins.

In the rocks of the Ilmenogorskii complex are localized numerous bodies of pegmatites of various composition and age (Fig. 2.6). The majority of researchers distinguish in terms of composition two main groups among them: miaskitic and granitic [18,23]. The intermediate position is occupied by syenitic pegmatites. Pegmatites of each of the groups are somewhat distinguished in terms of composition and texture from the corresponding types of magmatic rocks. Traditionally, to the granitic belonged the pegmatites containing primary rock-forming quartz, to the miaskitic belonged the nepheline-containing pegmatites, and to the syenitic belonged the alkaline feldspar pegmatites without quartz and nepheline.

The study of the regularity of formation of pegmatites of the Il'menskie Mountains conducted in recent years by V.I. Popova with co-authors has complemented and generalized the data on the composition and age correlation of vein bodies. In the plotting of over 100 cross-cuttings of various veins in 60 mines, they obtained the following delineation of groups of vein formations of various age—pegmatites (from ancient to young): pre-miaskitic granitic; feldspathic, miaskitic, and corundum-feldspathic; post-miaskitic granitic; and amazonitic and their analogous granitic.

The long-term history of formation of the vein field of the Il'menskie Mountains (from 430 to 180 Ma) in a complex geological block gave rise to the significant

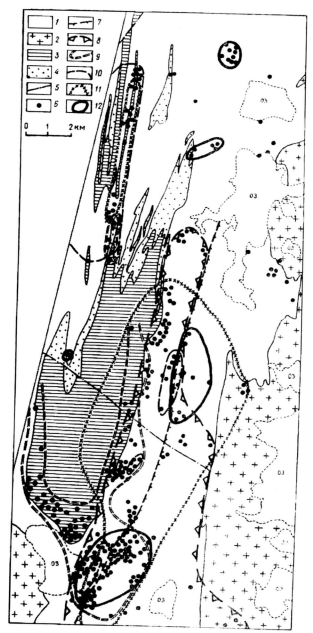

FIGURE 2.6

Schema of geological structure of the Il'menskie Mountains and the basic districts of development of amazonitic pegmatites: (1) metamorphic rocks (gneisses, amphibolites, migmatites, schists, and others); (2) granitoids; (3) miaskites; (4) syenites; (5) tectonic dislocations; (6) mines; (7–12) districts of development of pegmatites (7: granitic pre-miaskitic, 8: feldspathic, 9: miaskitic, 10: corundum-feldspathic, 11: granitic post-miaskitic, 12: granitic amazonitic).

variation of composition and texture of the pegmatites. The composition of pegmatites shifts sequentially from granitic to syenitic and later again to granitic, which provides evidence for the occurrence of genetically differing processes of mineralogenesis. For the main rock-forming minerals of pegmatites—feldspars—is noted the following evolutionary tendency. In the ancient granitic pegmatites, acidic plagioclases predominate; in the feldspathic, miaskitic, and corundum-feldspathic pegmatites are developed plagioclases, microclines, and anorthoclases; in the young granitic pegmatites, a principal role is played by anorthoclases and microclines, and in youngest amazonitic pegmatites by microclines. Also shifting in this regard are the composition and number of accessory minerals. Characteristic for the ancient granitic pegmatites is a poor accessory mineralization (betaphite, orthite, titanite, apatite, rutile, and others) of rare-earth specialization, for alkaline and young granitic, it is rare-metal and rare-earth (zircon, pyrochlore, columbite, samarskite, aeschynite, orthite, thorite, tscheffkinite, etc.), and for amazonitic and analogous pegmatites, it is primarily rare-metal (columbite-tantalite, beryl, phenakite, helvite, cassiterite, etc.).

The analysis indicates that almost all pegmatites were formed under the conditions of amphibolite facies of metamorphism of enclosing rocks, and only in the very end with the formation of amazonitic and similar pegmatites is noted a fall in temperature before the facies of green schists with local development of corresponding paragenesis. The process of formation of amazonitic pegmatites is replaced in time by local greisenization.

Amazonitic pegmatites of the Il'menskie Mountains. At the present time in the Il'menskie Mountains, 65 mines are numbered in the veins of amazonitic pegmatites. They are situated on the eastern slope of the Main Il'menskii ridge, where several groups are formed in meridionally elongated bands, tracking at a distance of 1 to 3 km from the miaskitic pluton. Relatively recently, veins of amazonitic pegmatites were discovered as well on the western slope of the Il'menskie Mountains. Several veins are localized in the fenites of the contact aureole of the miaskitic pluton. Granitogneisses and amphibolites commonly serve as enclosing rocks of amazonitic pegmatites (Fig. 2.7).

The form and mode of occurrence of veins of amazonitic pegmatites is fairly diverse. Of the 65 veins, 30 have lateral bearing, 13 have meridional and submeridional bearing, and 22 have diagonal bearing. Their fall is usually steep, close to vertical. The most widely distributed are thin (approximately 0.5 m), elongated (up to 150 to 200 m) bodies with short thick (up to 4 to 5 m) bulge or without them. The bulges are situated either in the knee-folds of thin bodies (veins 58, 60, 27) or in echelon folds in parallel veins (61, 100). From the bulge fairly often appear several apophysis (veins 61, 35, 57, 77). Occasionally occurring are short thick veins with bulges (395) and branching in sections of pinching out of bulges (veins 70, 3, 112).

The inner texture of veins of pegmatites is extremely diverse. In terms of composition and texture of the outer zones, pegmatite bodies can be divided into

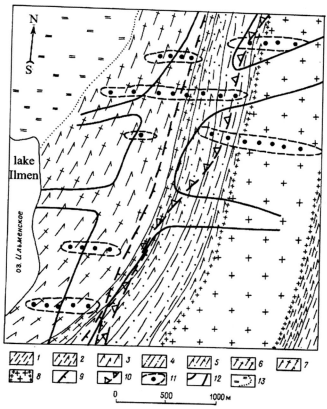

FIGURE 2.7

Schematic position of zones of development of variously aged vein formations in the southern part of the Il'menskie Mountains. (1–2) rocks of Argayach series (1: plagiogneisses, 2: alternation of amphibolites and plagiogneisses); (3–7) rocks of Ilmenogorskii suite (3: amphibolites, 4: quartzites with intercalations of plagiogneisses, amphibolites with intercalations of quartzites and of gneisses, 6: plagiogneisses, 7: amphibolites with intercalations of plagiogneisses); (8) migmatites and zones of near-contact main development of vein granites; (9–12) zones of development of pegmatites (9: pre-miaskitic granitic [western border], 10: feldspathic [eastern border], 11: miaskitic, 12: granitic amazonitic); (13) swamp.

three types: (1) quartz-microclinic graphic; (2) quartz-bifeldspathic zoned-graphic; (3) quartz-microclinic with zones of nonuniformly granular granite (Fig. 2.8).

Predominating are veins of amazonite pegmatites belonging to the first type in the quantitative respect. More often than not, they have the following zonality (from periphery to center): coarse-graphic non-amazonitic pegmatite—fine-graphic and apographic amazonitic pegmatite with gradual increase of quartz grains (Photo 2). A central part of the vein is occupied either by a quartz nucleus or by a cavity.

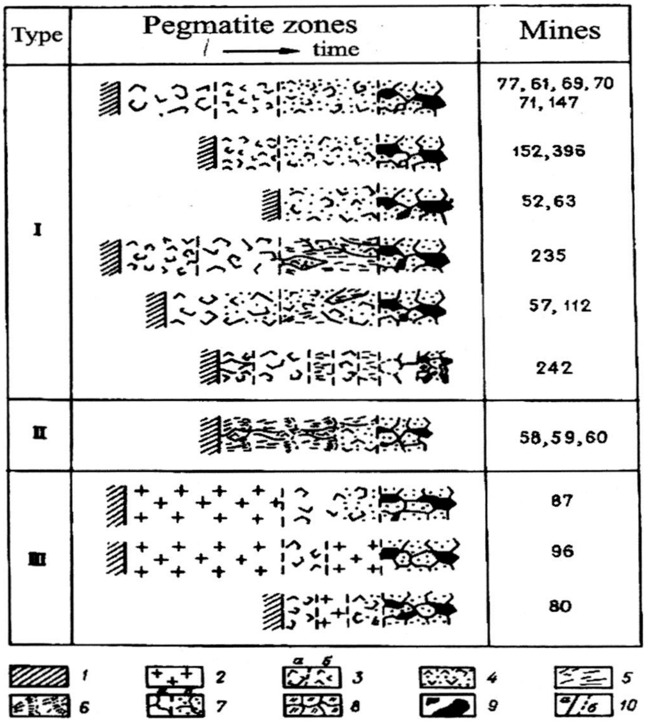

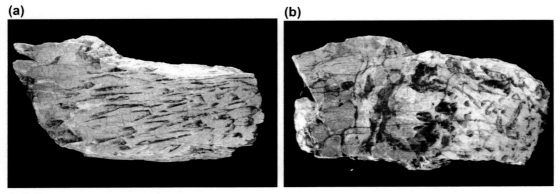

FIGURE 2.8
Generalized schemes of the sequence of formational zones in amazonitic pegmatites of the Il'menskie Mountains (from selvage to cavities): I–III: types of pegmatites (see text). (1) enclosing rocks; (2) nonuniformly granular borders; (3–7) pegmatite (3: quartz-microclinic [a: large-graphic, b: close-graphic], 4: quartz-amazonitic apographic, 5: quartz-albitic graphic, 6: the same with rhythmic-zoned distribution of quartz, 7: coarse-block [a: microclinic, b: amazonitic]); (8) coarse-block outcrops of albite-oligoclase; (9) quartz; (10) borders of the vein (a) and of the structural-mineral zones (b).

(a)   **(b)**

PHOTO 2
Development of the process of amazonitization in various textural zones of pegmatite bodies (Il'menskie Mountains, mines 70, 74): (a) from the graphic zone (yellow microcline + quartz) toward the apographic zone (amazonite + quartz); (b) from the apographic zone (quartz + potassium feldspar) toward the pegmatoid and blocky zone (amazonite + albite + quartz).

In the amazonitic pegmatites are manifest over 70 minerals, of which approximately 30 have been found only in recent years [47]. The primary minerals of the veins are quartz, microcline including amazonite, and albite; the secondary minerals are biotite, muscovite, garnet, topaz, and magnetite. We emphasize that many of the rare-earth and rare-metal minerals, previously considered to be characteristic for non-amazonitic bodies, at the present time have been uncovered also in amazonite-containing pegmatite veins. In the latter have been identified as well alkaline pyroxenes and amphiboles, previously having been noted only in non-amazonitic pegmatites. Typical accessory minerals of amazonitic pegmatites include the minerals of: Pb (galena, cosalite, Pb-pyrochlore), Be (beryl, phenakite, helvite, genthelvite), Bi (bismuthine, cosalite, bismuthite, clinobisvanite, lillianite, bismite), Sn (cassiterite, ixiolite), Ta (tantalite, microlite), and certain aluminofluorides (cryolite, weberite, etc.).

We pause on the characteristics of typical veins of amazonitic pegmatites. Mine 77 is the only one in the southern part of the reserve where the structure of the vein of amazonitic pegmatite in the thickest part of the bulge can be observed. The vein is embedded in the gneisso-amphibolithic series. The sweep of it is lateral, the fall to the north is at an 80° angle, and the thickness is approximately 2 m. In the western wall of the mine in the enclosing rocks is visible a thin vein of coarse-graphic pegmatite, being, apparently, the apophyse of the main vein. The texture of the vein in bulge is symmetrically zoned, characteristic for veins of the first type. In the mine are found quartz, microcline-amazonite, albite, biotite, garnet, topaz, fluorite, triplite, malacon, columbite, ilmenite, and magnetite.

The amazonitic pegmatite of Bliumovskaia mine (vein 50; Fig. 2.9) is embedded in the gneiss-amphibolite series. The vein has a thickness up to 8 m and is traced along the sweep from west to east at 150 m. The texture of the pegmatite body observed in the eastern wall of the mine is symmetrically zoned. The selvages are composed of coarse-graphic pegmatite, which closer to the center is replaced by fine-graphic. In the last zone occurs cavities with accessory beryl and topaz. In the central part of the vein is developed primarily non-quartz

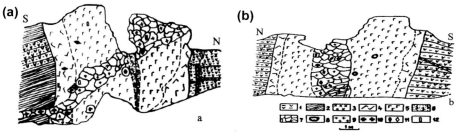

FIGURE 2.9
Cross-sections of the vein of amazonitic pegmatite of mine 50 according to Academic traverse (a: western wall, b: eastern wall): (1) granitogneisses; (2) amphibolites; (3–5) pegmatite (3: ancient granitic, 4: quartz-microclinic coarse-graphic, 5: the same, but fine-graphic); (6) quartz-albtite blastomylonite; (7) blocks of plagioclase; (8) cavity; (9) microcline; (10) amazonite; (11) biotite; (12) beryl.

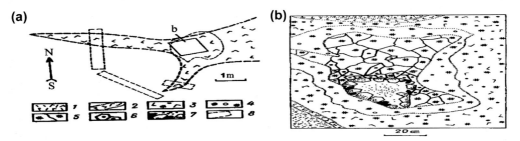

FIGURE 2.10
Form of the vein body of mine 235 (a) and detail of the structure of amazonitic pegmatite with cavity (b). (1–5) zones of pegmatite (1: quartz-microclinic fine-graphic, 2: the same, but coarse-graphic, 3: graphic and apographic quartz-amazonite-albitic, 4: pegmatoid and block quartz-albite-amazonitic, 5: block amazonitization); (6) crystals of biotite; (7) crystals of quartz (black) and amazonite (light) on the walls of the cavity; (8) contours of the mine excavation.

biotite-bifeldspathic block pegmatite with accessory samarskite, columbite, and monazite. Amazonitic color is observed in individuals of yellow microcline in the form of stains irregular and variable in form of pale-green color (with gradual transitions to sections of yellow color). The intensity and prevalence of amazonitic color gradually increases by measure of the transition from the selvages toward the center of the vein, and likewise from its upper parts to the lower. Alongside the basic mass of pale-colored yellowish-green microcline-amazonite are observed transverse small zones of intensively colored in greenish-blue color amazonite of thickness of approximately 1 cm, in which are included thin (several millimeters) veinlets of quartz.

A somewhat less widespread distribution is found for veins with the selvages composed of graphic non-amazonitic (mines 27, 396, 235, etc.) and of graphic amazonitic pegmatite (mine 63). The vein of amazonitic pegmatite of mine 235 forms a body isometric in plan in a section of the western branching of a thick lateral vein of fine-graphic pegmatite embedded in amphibolites (Fig. 2.10). Fine-graphic quartz-microclinic pegmatite of the peripheral zone is replaced closer to the center by a thin zone of coarse-graphic pegmatite. Further follows the zone of quartz-amazonite-albite apographic and graphic pegmatite, which either transitions into pegmatoid and block (amazonite from early to late zones gradually acquires saturated color), or borders with a cavity (see Fig. 2.10). A cavity of size 30–40 cm has a form close to isometric. On its walls are developed crystals of amazonite and quartz, overlaid by crystals of tourmaline and phenakite. Besides these minerals, in the cavity were uncovered muscovite, topaz, columbite, cassiterite, zircon, ixiolite, and microlite.

In several veins, separate inner zones are composed of bifeldspathic graphic pegmatite (mines 239 and 395). A lenticular vein of pegmatite of mine 395 with thickness up to 4 m and elongation of 25 m is embedded in amphibolites (Fig. 2.11). The zonality of the vein is expressed in the replacement of near-selvages quartz-microclinic pegmatite of megagraphic texture by bifeldspathic fine-graphic, which gradually coarsens and is replaced further by

(a) (b)

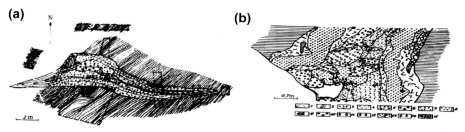

FIGURE 2.11

Schema of the structure of the vein of amazonitic pegmatite of mine 395: (a) plan, (b) cross-section of the eastern wall of the mine. (1) amphibolites; (2) ancient granitic pegmatites and aplites; (3–7) zones of pegmatite (3: coarse-graphic quartz-microclinic, 4: fine-graphic quartz-bifeldspathic, 5: coarse-graphic quartz-amazonitic, 6: block amazonitic, 7: graphic quartz-albitic); (8) albite (cleavelandite); (9) quartz nucleus; (10) topaz; (11) beryl; (12) biotite; (13) quartz-albitite mylonite; (14) muscovite; (15) block microclinic pegmatite.

quartz-amazonitic graphic. The latter in its turn, with coarsening and the subsequent disappearance of quartz ichthyoglypts, transitions to block amazonitic pegmatite. Idiomorphic terminal points of the crystals of amazonite are buried in the quartz of the nucleus. In quartz, together with the crystals of amazonite, occur crystals of topaz, golden-reddish-brown muscovite, and albite of comb structure. In the outer zones of growth of crystals of amazonite and topaz are noted accessory microlite and cassiterite. On the lower border of the quartz nucleus are situated small cavities with crystals of accessory beryl and topaz. In the form of incrustations in quartz have been identified galena, helvite, and pyrite. In the outer and inner zones of pegmatite are observed garnet and accessory columbite-tantalite.

Alongside the most widespread pegmatite bodies described above, others are known with highly unique structure. The brightest distinguishing sign of these veins is the presence and predominance in them of quartz-albitite pegmatite of radiating-corona rhythmic graphic texture (mines 35, 58, 59, 60, and others). Similar pegmatite is developed directly at the selvages, where among frayed individuals of albite with rhythms of quartz ichthyoglypts are located individuals of pale-colored amazonite, in which are traced the same rhythms, but of somewhat greater thickness.

Finally, altogether in several veins (mines 79–88 and 96) the essential part of the volume is occupied by nonuniformly grained pegmatite of granitic texture. In several mines, it is localized directly at the selvages of the veins; in others, it is preceded by a large near-selvages zone of quartz-microcline graphic pegmatite. Most frequently, such pegmatite of granitic texture directly transitions toward cavities with amazonite and its corresponding accessory mineralization: topaz, phenakite, tourmaline, and struverite.

It is necessary to mention that among pegmatite veins classified as non-amazonitic, many bodies in terms of structure as much as of mineral composition, and occasionally also in terms of age indicator, are practically indistinguishable from veins with amazonite. Such, for example, are quartz-feldspathic rare-earth pegmatites, which are embedded in pyroxene-amphibolite granite-gneisses and syenites and situated closer to miaskitic rocks than are amazonitic pegmatites. These circumstances, in the opinion of the author et al. [47], allow, first, to hypothesize the potential discovery of amazonite in several such veins in the course of further search-prospecting fieldwork, and second, to consider that the presence or absence of amazonitic color in potassium feldspars is not in the present case a sufficient criterion for typification of pegmatite veins. At the same time, thanks to the age setting, to the close internal structure, and to the similarity in the sequence of mineral-formation, amazonite-containing pegmatites are sharply differentiated into a single group of vein bodies.

In conclusion, it is necessary to refer to the fairly close mineral composition of Kola and Il'menskie pegmatites. Already in those as well as in others have been identified over 70 mineral species, a large part of which are represented by the accessory minerals tantalum, niobium, yttrium, beryllium, tin, lead, and bismuth, and a range of other elements (Table 2.4). It is highly indicative that several rare-earth and rare-metal minerals of these pegmatites possess almost identical chemical composition. These circumstances, as well as the proximity of geological setting and particularities of the structure of the Kola and Il'menskie pegmatites with rare-earth and rare-metal mineralization, enables the hypothesis that the conditions of their formation were fairly similar. Detailed analysis of these noted facts falls outside the framework of the present work, but nevertheless necessitates the following remarks.

The comparison of the geological settings of the western Keivy and the Il'menskie pegmatite fields provides evidence for several of their similarities—principally in terms of the presence in these districts of the similarly composed series of intrusion rocks with the most development among them of formations of alkali type. Thus, it turns out that almost all alkali rocks of one region have their analogs in the other—they are different aged homological representatives of alkaline-granitic and nepheline-syenite (miaskite) formational types. Homotypes among them are biotite nepheline syenites (miaskites), syenites and their pegmatites, as well as all types of alkali metasomatites.

Taking account of the aforementioned, the existence can be surmised of several other common features in geological setting of the analyzed pegmatites of the fields of the alkaline-granite formation. Confirmation is provided, for example, by the data of M.G. Isakov on the characteristics of amazonitic pegmatites of the Vishnevye Mountains, which according to the geological conditions of the finding are analogous to the Il'menskie amazonite-containing pegmatites. Here were described pegmatites with amazonite, situated at varying distances from alkaline rocks. Moreover, it was discovered that in the pegmatite vein located at the maximum distance from miaskites and syenites (by several kilometers), potassium feldspar is intensively colored in a turquoise amazonitic color, while highly weak amazonitization is characteristic for the pegmatite body embedded in the contact of granite-gneisses with alkali rocks.

Table 2.4	Comparative Characteristics of Typomorphic Accessory Minerals in Amazonitic Pegmatites of the Kola Peninsula (1) and the Urals (2)	
	Mineral phases	
Characteristic element	**1**	**2**
Tantalum, niobium	Plumbum-microlite	Columbite-tantalite,
	Plumbum-pyrochlore	Pyrochlore-microlite
	Microlite	Ilmenorutile-struverite
	Manganese-columbite	–
	Complex oxides	Complex oxides
Tin	Cassiterite	Cassiterite
	Wodginite	–
Bismuth	Native bismuth	Native bismuth
	Bismuth oxides	Bismuthinite
	Zavaritskite	
Lead	Galena	Galena
	–	Sulfosalts Pb and Bi
Fluorine	Yttrofluorite	Fluorite
	Fluorosilicates	Topaz
		Aluminofluorides
Rubidium	Microcline	Microcline
	Amazonite	Amazonite
Lithium	Zinnwaldite	Zinnwaldite
Beryllium	Gadolinite	Beryl
	Gingganite	Phenakite
	Genthelvite	Helvine
	Danalite	Danalite
Phosphorus	Phosphates	Apatite
Yttrium and rare earths	Silicates	Oxides
	Phosphates	Silicates
	Carbonates	Phosphates
	Oxides	
	Halides	

As has been noted, in the Kola pegmatites, the process of amazonitization reaches a maximum in vein bodies situated at a significant distance from alkali granites. And in the Il'menskie Mountains is noted a regular positioning of pegmatite veins with different intensity of the amazonitization process relative to the miaskitic pluton (toward the district of development of this pluton, in the opinion of the author et al. [47], spatially extends alkali granites still unexposed, and source for amazonite-containing pegmatites). Needless to say, the stated proposal requires careful verification. Despite the belonging of amazonitic pegmatites of the Kola Peninsula and the Urals to the alkaline-granite formation and in spite of the range of common features of their geology, there are also definite differences between them. The principal difference consists of the uniqueness of internal structure and composition of rare-earth/rare-metal mineralization.

At present, as the most general reasons for such differences should be taken the differing depths and age of formation of the granitoids that produce these pegmatites, as well as the differing position of the latter as regards the source alkaline-granite plutons.

2.4.3   Kazakhstan and Kyrgyzstan

The well-known intrusions of the subalkaline-leucogranite formation in this region are associated in terms of age of origin with various tectonic–magmatic cycles. They are particularly widely manifested in association with late Caledonian and late Hercynian magmatism. Granites of this formation are distributed nonuniformly in various types of geotectogenes of linear type [1].

The amazonite-containing granites of Kazakhstan occur in connection with granitoids of the late stages of tectono-magmatic cycles, and form at the end stages of the establishment of intrusion complexes in the form of phase and facies formations. They constitute small transverse intrusion bodies, not rarely of the fracture type (dikes, flat-lying deposits, a dome of larger granitoid plutons), and are characterized by polymineral composition, consistently wide development of lath-like albite, and rare-metal (columbite, cassiterite, zircon, etc.) and fluorine-containing (fluorite, topaz) accessory minerals. We will examine in more detail the structure and composition of several of these plutons, based on the data of our own research and of a range of other works [2,4,47] conducted by various geologists (Shcherba et al., A.N. Plamenevskaia, A.N. Bugaitsa).

The **Maikul' pluton** is situated in the zone of the intersection of the Chu-Ili anticlinorium and western Balkhash inner depression (Chu-Ili geotectogene). It has a fairly significant size (approximately $320\,km^2$) and, judging by the elongation in the lateral direction, coincides with the system of faults transecting the Sarytum zone, being in its own turn a branch of the Jalair-Naiman fault zone.

The pluton is composed of granites with radiological age of 300 to 400 Ma. Occurring in it are medium-grained and porphyroform biotitic, leucocratic, and amazonitic granites. The latter are developed predominantly in the western endocontact part of the pluton (and particularly widely developed in the district of the so-called Blue Peak), as well as in the eastern and south-eastern parts of the pluton (Fig. 2.12). The amazonitic granites consist of potassium feldspar (from 30% to 40%), albite (30–50%), quartz (20–30%), and zinnwaldite (1–3%). Of the accessory minerals here, occurring in significant distribution are fluorite, columbite, cassiterite, monazite, and topaz. The western contact of the intrusion is very gentle—it falls at an angle of 10 to 15° to the southwest to quartzite-type sandstones of the Precambrian, which in the contact with granites is thermally metamorfohosed, occasionally weakly greisenized. In the immediate endocontact is observed a deposit (approximately 2 m) of white with rarely blue grains of fine- to very fine-grained aplito-type granites with biotite. Under them are deposited medium-grained amazonitic granites (Photo 3). In the upper part of the latter (0.5 m) is traced nonuniformly grained rock with separate large "pisolitic" grains of quartz.

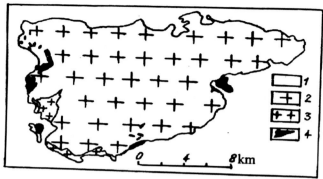

FIGURE 2.12
Schema of the positioning of amazonitic granites in Maikul' pluton. According to N.L. Plamenevskaia:
(1) enclosing rocks; (2–4) granites (2–3: medium-grained [2: porphyritic two-mica, 3: biotitic], 4:
amazonite-albititic).

PHOTO 3
Fine (a) and medium (b) -irregular-grained albite-amazonitic granites of the Maikul' pluton.

As the result of detailed study, the following was determined: (1) amazonitic granites, as well as rocks similar and associated with them (albitic and amazonitic aplites, albitites, amazonitic pegmatites, fluorite-quartz veins, and greisens) are not facies or derivatives of common biotitic and alaskitic granites of the pluton, and they represent the independent, youngest intrusion complex; (2) the manifestations of amazonitic granites and their genetic types are situated not along the periphery or apical of the Maikul' pluton, but within a narrow (1–5 m) meridional belt, transecting the western extremity of the pluton and extending for many kilometers to the north and south; (3) on the whole, the described pluton is layered by granites of three complexes: granitic (Kaibskii), leucogranite-alaskite, and subalkaline-amazonite-granite (western Maikul'). Evident also is the similarity of granite series of the given middle Paleozoic pluton and Mesozoic pluton of Siberia (see below).

According to G.N. Shcherba and collaborators, to the western Maikul' complex also belongs the small Kengkiik pluton situated to the north of Maikul' composed of biotite-albitic granites, in the pegmatite formations of which is noted blue amazonite.

In the north of the Chu-Ili geotectogene amazonite-albitic granites are found at deep horizons (400–500 m) of the Kara-Oba tungsten deposit. Here, evidently, occurs a superposition of amazonite-albitic granites of the subalkaline-leucogranite formation onto granites of the alaskite formation, with which is associated tungsten mineralization.

The **Kurmenty pluton** is situated in the district of the Kungei-Alatau Range (northern Tian Shan), in the upper Taldy River. It is part of the Kungei granitic pluton, putatively assigned by G.P. Lugovskii and other researchers to the late Caledonian intrusion complex. The earliest formation is represented by the adamellites and normal granites that constitute the large Kungei pluton. Later form the leucocratic and alaskitic biotitic granites, which constitute particularly the Kurmenty and other small plutons. The Kurmenty pluton occupies an area of 40 to 50 km^2 and controls the sublateral tectonic dislocation of the second order, marking the limit of the large structural blocks of the region. In the exocontact of the pluton is noted a thin aureole of greisenization. Among the rocks of the backs of the Kurmenty pluton is situated a series of thin vein bodies of granite-, aplite-pegmatites, and pegmatites. Many bodies of granite-pegmatites represent apophyses of the basic body of granites. Fine bodies of pegmatites occur also among granites of the peripheral zones of Kurmenty pluton and (more rarely) in the more deep-seated parts of the latter.

For the pluton, revealed by the stripping of the relief at a depth of approximately 500 m, is manifested the following vertical zonality. In the deepest of its parts is deposited light-gray and pinkish-red leucocratic biotitic granites, by chemical composition close to alaskites, but characterized by a somewhat elevated alkalinity (9% $K_2O + Na_2O$). To the main minerals of the granites belong albite (32–33%), potassium-sodium feldspar (30–31%), and quartz (33–34%). The content of biotite reaches up to 3 to 4%. The paragenesis of accessory minerals is as follows: magnetite, ilmenite (in small quantities), monazite, zircon, thorite, and fluorite (predominantly violet). Present as well are samarskite, fergusonite, and columbite.

Higher along the cross-section, according to G.P. Lugovskii et al., is observed a gradual change in the texture of the granites, expressed primarily in the development of a fine-grained aggregate of tabular albite of the second generation, the strengthening of idiomorphism of potassium feldspar and quartz, and finally the manifestation of porphyroblasts of the latter, the manifestation and the growth of the role of lithium mica and topaz, cassiterite, samarskite, and the relative elevation of contents of tantalum, tin, rubidium, fluorine, and other rare elements. In the same direction, beginning from a depth of the order of 150 to 200 m, is noted a gradual shift of color of potassium feldspar from common gray or pink to blue amazonitic.

The upper zone of the Kurmenty pluton of thickness of 30 to 50 m is formed almost exclusively of amazonitic granite, having the following composition:

albite (40–58%, which includes 30–55% lath-like of second generation), amazonite (17–27%), quartz (20–27%), zinnwaldite and relicts of protolithionite (together 2–3%), topaz (1–3%), as well as columbite-tantalite, black iron bipyramidal cassiterite of the second generation, malacon with elevated content of hafnium (and tantalum), colorless fluorite with elevated content of rare earths, strongly altered fluocerite (replaced by aggregate of bastnesite and fluorite), and magnetite of second generation (including pseudomorphoses of mushketovite after hematite).

The content of rubidium in the light-gray granites of the inner parts of the pluton comprises in the central 0.036% (with variations from 0.027% to 0.044%); higher along the cross-section in the weakly amazonitized granites it reaches 0.06–0.08%, while in the amazonitic granites of the peripheral part of the pluton, it is 0.13% (with fluctuations from 0.09% to 0.19%). The accumulation factor, that is, the relationship of rubidium content in the granites of the peripheral part of the pluton to the initial content in the unaltered granites of the deep zones, consists of 3.5 to 4. Simultaneously upward along the cross-section is observed an increase of lithium content from 0.0043% to 0.024% (accumulation factor of 5 to 6).

Granite-pegmatites and pegmatites embedded among the granites of the upper zone and in the supra-intrusion zone of the Kurmenty pluton, among rocks of the back, in terms of a range of particularities of mineral and chemical composition, intensity of albitization and greisenization, and other indicators can, probably, be applied to the upper fifth zone of the pluton. Potassium feldspar in them is represented principally by highly intensively colored (amazonitized) maximum ordered microcline; albite of the second generation sharply predominates over the early albite, quartz content does not exceed 25–30%, and among micas are developed light-green zinnwaldite and the almost colorless cryophillite replacing it. The sizes of the crystals of yellowish topaz in the pegmatites reaches 1–1.5 cm.

The opinion of G.P. Lugovskii and others that the zonality of the structure of the Kurmenty pluton is determined by the gradual autometasomatic change of the initial leucocratic and alaskite granites of the second phase is not shared by G.N. Shcherba and collaborators, who consider amazonitic and albitic granites in the capacity of the youngest Mesozoic formations (analogous to the amazonitic granites of Maikul' pluton), superposition on Caledonian leucogranite-alaskites. The author et al. [47] support the latter conclusion and consider the amazonitic granites of the Kurmenty pluton as typical representatives of the subalkaline-leucogranite formation.

The **Karagaily-Aktas pluton** is situated on the eastern periphery of the northern Tian Shan geotectogene and structurally coincides with the graben-synclinal block composed of Cambrian marmorized limestones, dolomites, and schists.

The pluton represents a large dike body of sublateral sweep, elongated with interruptions almost for 9 km, having a thickness of 10 to 400 m. The contacts of the intrusion are steep (60–80) in the northern part and gentler (30–60) in the southern. Granites of the pluton belong to the amazonite-albitic type. These are the albitic granites with weak and nonuniform amazonitization, the intensity of which increases with depth. Amazonite-albitic varieties are more widely developed in the central part of the dike. On the flanks amazonitization tapers out. The granites often are greisenized and transected by a network of quartz, quartz-feldspathic, quartz-mica, and mica-feldspathic veins and veinlets with accessory (Ta-Nb) mineralization. In the exocontact of the dikes, the enclosing rocks are skarnitized and greisenized, containing ore veins and veinlets.

Albitic granites of the pluton are rocks of close- and medium-grained (sometimes porphyritic) texture, consisting of quartz (30%), microcline including amazonite (15–30%), lath-like albite (up to 30 to 50%), zinnwaldite (up to 5%), and muscovitized biotite. Among accessory minerals are distributed fluorite, topaz, sellaite, columbite, cassiterite, wolframite, zircon, thorite, etc. The tantalum and niobium most typical for these granites along with columbite-tantalite are concentrated in the main minerals of rare-metal ores: wolframites and cassiterites. According to the data of various researchers, characteristic for the intrusion is a distinct vertical petrogeochemical zonality.

There is no single opinion regarding the age of the granites of the Karagaily-Aktas pluton. Some researchers consider them Paleozoic (Caledonian), others Mesozoic.

Besides the above-described Kurmenty and Karagaily-Aktas plutons, in the northern Tian Shan geotectogene, still other plutons of amazonitic granites are known: Koturginskii, Talzhanskii (Kungei-Alatau Range), Aksaiskii, Tuiukuiskii, and Tastynskii (Zailiiskii Alatau), as well as the Dzhulysuiskii, Keregetashskii, and Ton-Tuiuktorskii (Terskei Alatau Range). For all or for the majority of these complex plutons is proposed a late Caledonian age of formation of the early phases of granitoids and a Mesozoic age of activity, in the period of which also arose the phases of amazonitic pegmatites.

Khorgos pluton is situated on the southern slope of Jungar Alatau, between the rivers Chizhin in the west and Khorgos in the east. The first amazonitic granites of this district were described already in 1934 by A.K. Zherdenko, who performed a specialized survey of ore bodies of tin and other rare metals. The pluton by size not less than 10 by 4 km extends in a lateral direction, has a triangular form, branches out to the east, and coincides with one of the sub-lateral faults. In the north, it transects the larger Tyshkanskii pluton of biotite-hornblende granites of Carboniferous age, and in the south early Carboniferous volcanogenic rocks. The southern contact of the pluton, probably, is flat-lying, its separate outcroppings exposed in the south among acidic effusions and tuffs. The latter along the contact are hornfelsed, quartzified, and sometimes greisenized.

The pluton is composed primarily of biotitic porphyritic granites (sometimes of leucocratic face) of the first phase with widely developed marginal subfacies of granite-porphyries. These granites are broken through by stock- and dike-type elongated bodies of amazonitic porphyritic granites. According to the data of A.K. Zherdenko, a distribution band of amazonitic granites of a width of approximately 1.5 km extends from Chizhin in the north-east in the upper Karagaily Bulak stream and further along the right slope of Kaskabulak and the upper Arasan, crossing the River Khorgos into the territory of China, where they constitute the peak Oksantau (Syrtas). The largest bodies of amazonitic granite are situated within the boundaries of this band in the upper Karagaily Bulak, confined to the southern exo- and endocontact of the biotitic granites of the first phase. The thickness of the separate bodies fluctuates from 5 or 10 to 20 or 30 m, sizes from 0.1 to 0.5 km^2. The form of the bodies is rather variable; they often taper and form bulges up to several dozen meters. Such bodies are noted as well along the north-western contact of the pluton.

The contacts of the bodies of amazonitic granites with their enclosing biotitic granites and effusive rocks of the Lower Carboniferous are distinctly transverse. In the development area of the stock of amazonitic granites is exposed a fairly significant quantity of veins and veinlets of quartz-feldspar composition, and of pegmatites occasionally amazonitized to an insignificant degree.

Amazonitic granites are represented as almost white rocks with bluish or greenish hues of variously grained structure, consisting of bluish-green amazonite, white albite, gray semi-transparent quartz, and dark mica occasionally with golden lustre. The rock is made porphyritic by large (up to 5 or 6 mm) grains of amazonite and by the surrounding deposits of quartz of a size up to 2 or 3 mm. The matrix of rock is small-grained aplitic. Porphyritic phenocrysts comprise from 30% to 80% of the general mass. The cemented mass is formed by albite (50–60%), quartz (25–35%), potassium feldspar (10–15%), and biotite and muscovite (together no more than 5%).

As it almost always does, amazonite has a microclinic cross-hatched twinning; in it are practically absent perthitic incrustations of albite and perthitic material, but there is present a fair amount of secondary albite. The latter in places constitutes almost monomineral veins.

Accessory minerals of amazonitic granites are zircon, magnetite, titanite, orthite, topaz, and fluorite. In the dike bodies in places, granites are converted into fluorite-mica-topaz greisens.

In the endocontacts of amazonitic granites are observed pegmatoid sections of size from 1 or 2 cm up to 1 m. In independent vein-form bodies of amazonitic pegmatites besides amazonite, quartz, and muscovite is noted a deposit of columbite, molybdenite, bismuthine, and other minerals.

In the **Jungar** geotectogene, besides the described Khorgos transverse body of amazonitic granites and pegmatites, occurring in the **Oysaz pluton**, there are biotitic and biotite-hornblende granites. The late Hercynian plutons with younger amazonitic granites also include the **plutons Dogolan** (dikes and flat-lying deposits of amazonite-albitic granites) and **Kyngyrzhal'** in the Chingiz-Tarbagatai geotectogene; the late Caledonian **Zolotonosh** pluton (with exocontact zones and apophyses of amazonite-albitic granites) is noted in the Kokshetau geotectogene.

2.4.4 Eastern Siberia

In the late 1950s and early 1960s, on the territory of eastern Siberia was discovered a large province of rare-metal deposits connected with albitized and greisenized (including amazonitic) granites. According to A.A. Beus, A.I. Ginzburg, L.G. Fel'dman, and other researchers [2,47], rare-metal granitoids of lithionite-microcline-albitic composition (subalkaline-leucogranitic in the classification and terminology accepted by the author of the present work) always are late orogenic and are situated in typically geosynclinal districts or are associated with zones of activity.

The plutons of rare-metal granitoids are commonly represented by fracture bodies, steep-falling stocks, and small domes often having zoned structure formed in the beginning of the late Jurassic epoch at small depth. In this region, the overwhelming majority of rare-metal granitoids are embedded in sedimentary-metamorphic rocks: sandstones, schists, and aleurolites, conglomerates of Jurassic age in which the presence of aureoles of near-intrusion alterations have been identified (the development of mica, topaz, and cassiterite). We stop here on the characteristics of a range of typical plutons of Mesozoic granites.

The first Mesozoic pluton (Orlovskii deposit) of eastern Siberia [4,47] is situated in the field of metamorphosed sandstone-schists rocks of the upper Proterozoic to Lower Cambrian (Fig. 2.13). It consists of three closely neighboring granitic plutons: eastern, central, and western, by area respectively 0.06, 8, and 1.8 km^2.

Up to 80% of the central pluton from the surface is composed of granites of two varieties. Predominating are medium-grained nonuniformly grained porphyritic biotitic and two-mica granites with the size of the grains in the matrix of 0.5 to 3 mm. In a secondary quantity occur fine-to medium-grained rare porphyritic and highly nonuniformly grained biotitic and two-mica granites. Medium-grained porphyritic granites are broken through with rare dikes of fine-grained vein granites. In the eastern extremity of the central pluton are observed alaskites.

In the eastern pluton (not shown in Fig. 2.13) on the surface are exposed principally fine-grained greisenized alaskites with more or less large xenoliths of medium-grained porphyritic granites.

The western pluton has in plan an irregular isometric form (see Fig. 2.13). The surface of the western contact at an angle of 20° is buried under the enclosing rocks, the eastern contact is steep (40–50°), the southern and northern contacts

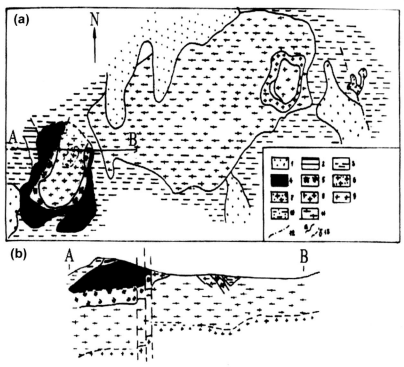

FIGURE 2.13

Schematical geological map (a) and vertical section (b) of Mesozoic pluton of Siberia (according to Beskin et al., 1979): (1) contemporary sediments; (2) upper Proterozoic–lower Cambrian metamorphosed sandstone-schists and effusive rocks; (3–4) complex of subalkaline granites (3: fine- to medium-grained amazonite-albitic with lithionite and muscovite, 4: mid- to large-grained muscovite-microcline-albtitic with *pisolitic* quartz); (5–6) complex of alaskites (5: medium-grained, 6: large- to medium-grained rare porphyritic); (7–8) complex of granites (7: fine- to medium-grained dense-porphyritic biotitic and two-mica, 8: medium-grained porphyritic biotitic and two-mica); (9) faults; (10) geological boundaries (a: relatively proven, b: proposed).

are still steeper (50–70°); to this they, particularly the northern, are bound by displacement along the faults. The pluton has an asymmetrical concentrated-zoned structure: in the center are embedded two-mica porphyritic granites, from the outer side is observed a crescent-shaped, exposed to the north outcrop of albitic granites. With more detailed examination, it is revealed that from the center to the periphery or in the cross-section from bottom to top replacing each other are the following varieties of granites (Fig. 2.14): (a) biotitic, two-mica, muscovitic nonuniformly grained porphyritic (inside the crescent); (b) muscovite-microcline-albitic variously grained, with surrounding grains of quartz; (c) amazonite-albitic with muscovite and lithium mica fine-to medium-grained; (d) lepidolite-albitic with topaz fine-grained (under schist series). Of the other varieties, we note dikes of amazonite-albitic aplites, transecting two-mica porphyritic granites; amazonitic pegmatoid with zinnwaldite bodies embedded in

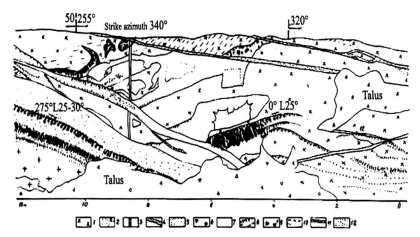

FIGURE 2.14
Variously grained rare-metal granites and vein formations in the south-eastern shelf of the horizon of
1000 m of the northern open pit of the western pluton: (1–3) granite (1: fine- to medium-grained amazonitic,
2: the same with zinnwaldite and stains of greisens containing quartz-amazonitic large-crystal zones,
3: fine-grained albitic with topaz); (4–5) quartz-albitic aggregate (4: fine-grained, 5: thin-banded);
(6) medium-grained quartz-amazonite-albitic granite with lepidolite, topaz, columbite-tantalite; (7) quartz;
(8) amazonite; (9) accessory beryl; (10) zinnwaldite, (11) quartz-fluoritic vein with molybdenite;
(12) banded aggregates of quartz with albite, beryl, topaz, zinnwaldite, and amazonite.

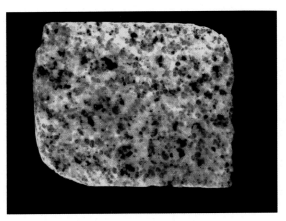

PHOTO 4
Medium-grained amazonite-albitic granite (Western pluton, eastern Siberia).

fine-grained banded albitites among albitic granites (Photos 4 and 5) that predominate
in the peripheral part of the pluton and are exposed in the exocontact,
the branching pluton of a dike of microcline-albitic aplite-type granites with
topaz. The near-contact alterations of schists are expressed in muscovitization
and quartzification; in the band with thickness of several meters are developed
lepidolite, fluorite, and sulfides.

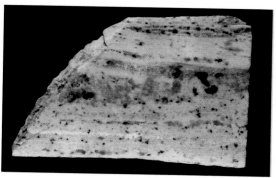

PHOTO 5
Coarse-crystalline quartz-amazonitic zone in banded albitite (Western pluton, eastern Siberia).

A deposit of lepidolite-albitic granites and the upper part a deposit of amazonite-albitic granites are tantalum-bearing ore bodies, where alongside tantalite, columbite-tantalite, and microcline occur garnet, fluorite, wolframite, cassiterite, rutile, and sphalerite. The main feature of the zonality is an increase from bottom to top of content of tantalum and of the ratio of Ta to Nb. The ratio of Na to K remains stable.

Contrary to the widespread opinion that albitic and amazonite-albitic granites of the western pluton are metasomatic rocks (apogranites) that formed according to ordinary porphyritic biotitic granites, the author et al. [47] have concluded on the independence of the complex of principally albitic granites.

Consequently, within the boundaries of the western pluton, we are concerned with two differently aged complexes of granites: the ancient porphyritic complex with vein, as well as the porphyritic varieties; and the young complex, the establishment of which began from the medium-grained albitic granites with *pisolitic* quartz, on top of them, possibly, was also superimposed late amazonitization.

The second Mesozoic pluton of eastern Siberia (Etyka deposit) initially drew the attention of researchers with regard to the tin-ore deposit; subsequently in the amazonitic granites was also discovered an industrial content of tantalum. The characteristics of the Etyka pluton have been supplied according to the data of G.P. Lugovskii et al., as well as of O.D. Levitskii et al., Iu.I Temnikov, and P.V. Koval'. The pluton is situated within the boundaries of the synclinal depression filled with the dislocated terrigenous rocks of early to middle Jurassic age. Among granitoids constituting the pluton are distinguished the following varieties: the earliest formations of the first phase exhibit bodies of granodiorites and quartz diorites, dikes of plagiogranitoporphyry, and others; the younger intrusions of porphyritic biotitic granites and the subsequent leucogranite-alaskites, as well as the small fracture intrusions of amazonitic granites.

One of the most fully researched plutons of amazonitic granites represents a small (1 by 1.2 km) body of asymmetrically domed form, slightly elongated in the submeridional direction. Its internal structure is fairly complex. According to P.V. Koval', amazonitic granites are associated with biotitic, leucogranitic, and alaskitic, predominantly porphyritic, granites. Fairly large bodies of such rocks either are directly in contact with some fine bodies of amazonitic granites or are located at some distance (up to 5 km) from them. In the first case, four bodies of amazonitic granites, situated in the endocontact of the intrusion of weakly porphyritic biotitic granites (on the border with porphyritic granodiorites and quartz diorites), are underlain by those same biotitic granites.

Taking into account that intersections by amazonitic granites of quartz-wolframite veinlets have been noted, which, in the opinion of the author et al. [47], ought to be associated with the leucogranite-alaskite complex, we can speak of an analogy of the structure of this Mesozoic pluton with the already examined first Mesozoic pluton. In this and in other cases, we are concerned not with phases (in the understanding of G.P. Lugovskii and collaborators), but with the different complexes of granites: the early complex, granitic without ore; the later leucogranite-alaskite complex with wolframite ore; and the final, youngest amazonite-albitic complex, to which is genetically associated tantalum (and tin) ore.

Following this scheme of formation of different aged complexes, we characterize rocks observed in the cross-section of the body of amazonitic granites.

The deep horizons of the pluton (350–550 m from the surface) are composed of homogenous uniformly medium-grained (grain size 2–3 mm) weakly albitized granites with very pale-colored amazonite. In a range of sections, the color of potassium feldspar is yellowish-pink. Along with microcline (18–26%), an essential role is played by albite (31–43%), quartz (24–32%), and lithium-ferruginous mica (2–8%). Accessory minerals are represented by fluorite, zircon, columbite, cassiterite, and sulfides. In the upper parts of the cross-section are noted rare fine grains of topaz. The hypidiomorphic-grained texture of the granite is only slightly veiled by weak albitization. In the pale-pink sections of the granite in the potassium-sodium feldspar are observed perthites of decomposition, in significant quantities occur coarse-perthitic feldspar, the microclinic cross-hatched twinning in which is expressed weakly or not markedly. Blue transparent K-feldspar-amazonite is commonly twinning or pseudo-twinning. From bottom to top along the cross-section in the rock grows the content of albite, the relative quantity of cross-hatched twinning microcline, and the intensity of blue color, protolithionite is replaced by zinnwaldite, and topaz appears. In the mass of albitized granite are developed thin veins and small zones of albitites, quartz-K-feldspathic and topaz-quartzic veinlets. On the whole, the examined rocks, apparently, may belong to one of the late leading phases of normal granites, slightly altered under the impact of the later-forming alaskites and amazonitic granites.

Higher along the cross-section (after the highly sharp border) up to the surface are traced medium- to coarse-grained albite-amazonitic granites (porphyroblast amazonitic intensively albitized granites of the main facies, according to P.V. Koval'). These rocks account for approximately 95% of the volume of the pluton

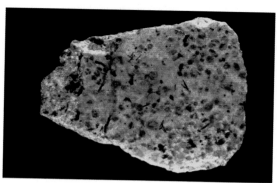

PHOTO 6
Medium- to coarse-grained albite-amazonitic granite (second Mesozoic pluton, eastern Siberia).

in its peripheral part. They predominate in the western, northern, and southern parts of the mapped surface of the pluton (Photo 6), containing more than 50% albite, from 15% to 25% quartz and potassium feldspar (predominantly amazonite), from 0.5% to 4% zinnwaldite, and up to 1% iron oxide. Characteristic for rocks are the following: relatively large size of grains of brightly colored bluish-green amazonite (from 2 to 3 to 4 mm); low content (5–14%) of fine-grained quartz and high content (25–40%) of the coarse porphyroblast quartz; the elevated (in comparison with other rocks) content of lithium-iron mica, in the form of large plates, the often taxitic texture caused by uneven distribution of tabular albite in the form of more or less large, almost monomineral stains, zones, and striae on the backdrop of the predominating mass of that same albite, evenly developed in the volume of the rock; and the manifestations of the exchange of microcline of checkered albite, occurring sporadically in the upper parts of the pluton and gradually acquiring greater significance in the deep horizons.

From bottom to top along the cross-section in the medium- to coarse-grained amazonitic granites is observed an increase in the quantity of albite, a strengthening in the intensity of amazonitic color of potassium feldspar, and a certain increase in content of lithium in mica. In terms of composition, the mica corresponds to zinnwaldite or to iron lepidolite. Moreover, the general content of mica in the rock decreases down to 1% or less, while the content of topaz grows. Changing, likewise, is the character of the accessory mineralization: instead of columbite, the leading mineral becomes pyrochlore-microlite. These particularities of the composition enable us to distinguish within the boundaries of the examined granites a lower, large zone in terms of volume with columbite and a significantly smaller upper zone with lepidolite and microlite. Similar granites are distinctly later in relation to all the remaining granites. In the authors' view, this is one of the leading first phases of the subalkaline-leucogranite formation.

Directly within the section adjoining the northern ridge-shaped outcropping, as well as in the more extensive area in the eastern half of the pluton under its eroded ridge, are distributed medium- to fine-grained amazonitic intensively albitized granites of the marginal facies (crust), according to P.V. Koval' (or domed facies, according to G.P. Lugovskii et al.). These are light pale-green, rarely grayish, even-grained rocks, in typical specimens consisting of 60% or more of lath-like albite. The content of quartz and amazonite in them reaches 20% (each). The quantity of topaz and lepidolite usually does not exceed 1%.

Within the boundaries of the pluton are observed a large quantity of hydrothermal veins. In the upper domed part of the pluton is noted a series of sharply falling dendroid zones of submeridional sweep, composed of fine-grained aplitotype albitized amazonitic granites. Characteristic for them are small sizes of grains of microcline and quartz (0.25–1 mm), an almost total absence of porphyroblast quartz, a low content (0.5–2%) of zinnwaldite, represented by fine flakes, and an uneven distribution of fine-tabular albite in the form of clusters, striae, and veinlets (its content consists of usually 20–30%).

Among the vein formations belong also the comparatively sparsely distributed medium- to coarse-grained vein amazonitic granites and granite-pegmatites. Typical for these granites are a highly weak albitization and a brightly expressed idiomorphism of amazonite with regard to quartz of the second generation, consisting of practically all the quartz of these rocks. The quartz-amazonitic veins occur not only in granites of the dome of the pluton, but likewise beyond its boundaries in the exocontact.

The vein amazonitic granites and granite-pegmatites are the youngest granitic rocks. We note that particular to them, precisely as to certain hydrothermalites of the exocontact, are the most intensive color of amazonite and the most ordered structure of potassium feldspar (maximum microcline). Alongside the amazonitic veins in the exocontact of the pluton are developed quartz-K-feldspathic veins with common colors of feldspar.

To the latest vein formations belong quartz veins with topaz and zinnwaldite, quartz veins having a micaceous margin, zones of quartzification with sulfides and singular calcite veins with galena.

2.4.5 Mongolia

Mesozoic granites, among which are fairly widespread plutons of rare-metal amazonite-containing granites, on the territory of Mongolia are distributed unevenly and situated exclusively in the eastern part of the country, toward the east from the meridian of 103° E. The area of their development controls unique block-wave structures of the type of systems of linear bending folds of the Premesozoic crystal formation, divided by depressions.

V.I. Kovalenko with collaborators distinguished within the boundaries of the examined region rare-metal granites of three geochemical types: (1) Modoto or standard according to the main geochemical indicators (primarily corresponding to the leucogranite-alaskite formation); (2) Sharakhadin, distinguished by lithium-fluoride specification (representative of the subalkaline-granite formation);

and (3) Dashibalbar with characteristic geochemical features of agpaite granites (the alkali-granite formation).

Granites of Modoto geochemical type constitute a large number of plutons, often multiphase. Particular to granites of the late phase is the presence of pegmatites of schlieren as well as of vein type. Directly behind the formation of granites of the late phase is noted a stage of formation of muscovitic greisens and quartz veins with the tin and tungsten ores typical for the region.

Among granites of the Sharakhadin type are known multiphase (Zhanchivlan, Yudugyin, Baga-Gazryn) and monophase (Abdar, Borun-Tsogtin) plutons. Their distinguishing particularity is the presence in the composition of the plutons of amazonitic granites or pegmatites. Commonly, granites of Sharakhadin type are coarse- or medium-grained alaskites with biotite, sometimes porphyritic, with surrounding morion quartz. By chemical composition they are close to alaskite and aplite from R.A. Daly (*Igneous Rocks and the Depths of the Earth*, 1933) which are characterized by the elevated by comparison with Clark contents of lithium (by 2–10 times), rubidium (by 2–10 times), and fluoride (by 2–20 times), and occasionally beryllium, zinc, tungsten, molybdenum and tin. The particularly high contents of these elements, as well as of niobium, tantalum, and hafnium are intrinsic principally to late rocks of the final intrusion phases and to rocks of the peripheral parts of the pluton. Common for them are lower values of ranges of potassium-rubidium (up to 20 or 40), of niobium-tantalum (up to one or less) and zirconium-hafnium (up to one) relationships. A highly typical geochemical feature of Sharakhadin granites are the anomalously low (by one to two orders lower than Clark) contents of barium and strontium.

From metasomatic formations connected with granites of the Sharakhadin type are developed zoned bodies from quartz-lepidolitic greisens up to monomineral albitites, as well as zoned bodies of zwitters, microclinites, and quartz-muscovitic greisens. All these rocks, similarly to granites of the final intrusion phases, are consistently ore-bearing with tin and tantalum and, in some cases, with tungsten, rare earths, and others.

The principal variety of granites of agpaitic or Dashibalbar geochemical type are monofeldspathic granites with riebeckite or aegirine. More rarely occurring are syenites and granosyenites. Frequently observed are the sharp connection and spatial proximity of these granites with their effusuve analogs. By chemical composition, these rocks belong to the typical alkaline granites with a magnitude of agpaite coefficient greater than 1 (sometimes significantly). In them are noted Clark contents as well as elevated (by one to two orders) contents of lithium, rubidium, tin, zinc, niobium, zirconium, rare earths, and thorium.

The temporal correlation of granites of various geochemical types may be judged according to the most studied Zhanchivlan pluton of complex structure, in which it was established that granites of the Gorikhin type (granitic formation) are broken through by those of Modoto type, and those in their turn are split by those of Sharakhadin type. Analogous temporal correspondences between granites of the enumerated types are manifested within the boundaries of the Tumen-Tsogtin pluton. In the Ongon-Khairkhan pluton have been observed intersections of quartz-wolframite veins connected with granites of Modoto type with a dike of topaz-containing quartz of keratophyre (ongonites) of Sharakhadin type. In the Baiian-Ulan pluton granites of Sharakhadin type are intersected by alkali granophyres (agpaitic type).

These data enable a hypothesis that within the boundaries of early and late Mesozoic intrusion cycles the sequence of formation at least of Gorikhin, Modoto, and Sharakhadin granites is similar and corresponds to the order of their enumeration. Evidently, the granites of agpaitic should be considered the latest in the early Mesozoic cycle. It is not improbable that their correlation with granites of the Sharakhadin type in separate plutons may be more complex.

As has been noted, amazonitic granites developed within the boundaries of Mongolia represent one of the varieties of Sharakhadin granites. According to mineral composition, these are amazonite-albitic granites, among which are distinguished varieties with zinnwaldite, as well as with topaz and zinnwaldite.

Amazonite-albitic granites with zinnwaldite (or biotite) are widely distributed in plutons of granites of the given geochemical type. In the medium-grained variety of these granites most frequently occurs cross-hatched twinning microcline (amazonite) without perthitic material, occasionally with relics of non-twinning, with a large quantity of perthitic ingrowths of potassium feldspar. In amazonites the quantity of perthites decreases to 10% by comparison with non-amazonitic potassium feldspars; the predominant form of them is scalloped. Among granites of the Abdar pluton are observed zoned crystals of potassium feldspar with non-twinning perthitic in the central part and amazonite along the periphery. Most frequently occurring in the medium-grained amazonite-albitic granites of the Abdar pluton is zinnwaldite; biotite is most frequent in the same granites of the Borun-Tsogtin pluton.

Amazonite-albitic granites with porphyritic texture and fine-grained granophyric matrix in distinction from medium-grained varieties, commonly contain fine pegmatoid schlieren. The peripheral part of them is composed of a granophyric aggregate of granite, and the central part by idiomorphic crystals, larger here than in the surrounding rock, of amazonite, quartz, and zinnwaldite. In the endocontacts of the Abdar and Borun-Tsogtin (eastern) pluton are manifested likewise large schlieren of amazonite-albitic pegmatites, consisting of the same minerals as amazonite-albitic granites.

In the apophyses and dikes of amazonite-albitic granites occasionally is observed a transition from alaskites with biotites to amazonite-albitic granites. In them occur confluent felsic varieties of amazonite-albitic granites. In the dike of Abdar pluton from the crescent to the back has been noted a gradual replacement of fine-grained alaskite with biotite by amazonite-albitic granite, and the latter's replacement in turn by amazonite-albitic coarse-grained pegmatite.

Moreover, the sections of amazonite-albitic granites occur in the central part of the pluton, in the peripheral parts of the domes of the second order.

Along the direction toward the contact, the amazonite-albitic granites become initially porphyritic at the expense of the manifestation of the micropegmatitic matrix, subsequently becoming fine-grained with a large quantity of fine schlieren pegmatoid sections. Directly at the contact, as a rule, is traced a zone of sharply unevenly grained rocks corresponding in terms of composition to aplite, with coarse and fine schlieren of quartz-microclinic pegmatites. Transitions between the described rocks are often gradual, but cases have been noted as well of the intersection by endocontact aplitotype and even by porphyritic amazonite-albitic granites of rocks of the central parts of the pluton.

In the eastern endocontact of the pluton has been observed a dike of aplitic and amazonitic granites, in which is repeated the zonality of the rocks of the entire pluton. Its middle visible thickness is approximately 5 m. The rocks of the hanging endocontact of the dike are represented by unevenly coarse-grained pegmatoid granites with deposits of amazonite and quartz by size in diameter up to 5 cm. The principally more fine-grained mass of the rock is composed of amazonite, albite, quartz, and bluish-green zinnwaldite. Along the direction toward the crescent of the dike, the rocks acquire an altogether more fine-grained texture, and the quantity of amazonite and mica in them decreases just as in the recumbent side appears aplitotype granite with biotite and common potassium feldspar in place of amazonite. With the decrease in thickness of the dike along its sweep to the northeast, the peripheral zone of pegmatoid amazonitic granites and the pre-crescent pegmatoid aplitotype granites with biotite taper out. Amazonitic granites acquire an altogether more fine-grained texture and transition into confluent amazonitic felsites with distinct endocontact zones of tempered and fluid textures.

The **Zhanchivlan pluton** represents an example of plutons of complex structure. Serving as enclosing rocks for granites of this pluton in the southwest are the sandstone-schists series of Proterozoic–lower Cambrian, in the west and in the north are deposits of the Devonian, and in the west are diorites and quartz diorites of Paleozoic age. The age of the granites of the Zhanchivlan pluton, according to the data obtained by the K-Ar method, fluctuates in interval of 183 to 231 Ma. The position of the pluton controls a large system of faults of northeastern sweep.

The basic backdrop of the Zhanchivlan pluton forms a large outcrop of sharply porphyritic granites of Gorikhin type. The area of their outcrop is more than 100 km^2. These rocks are broken through by medium-grained granites of the Modoto type. The latter are developed within the boundaries of several bodies in the northwest of the pluton with a general area of 20 km^2. Finally, to the youngest belong the rare-metal granites of Sharakhadin type, constituting several small outcrops of general area of approximately 50 km^2: Bural-Khangai, Ulan-Buridin, Urtu-Gotszogor, and others. Their contacts with the more ancient granites of Gorikhin and Modoto types are sharp and intrusive, with injections in the enclosing granites. The endocontact varieties acquire a sharply porphyritic aspect and not infrequently transition into granite-porphyries. Exocontact zones of the enclosing granites are somewhat rich in biotite, but do not change their granularity.

In the endocontact of the Sharakhadin granites in the Urtu-Gotszogor district have been observed magmatic breccia, in which fragments of crystals of quartz, potassium feldspar, and oligoclase are cemented by a felsitic quartz-feldspathic mass. In granites of the given type have been noted schlieren pegmatites with cavities filled in by large crystals of microcline (occasionally amazonite), albite, quartz, biotite, and tourmaline. The vein series of these granites is represented by granite-porphyries that resemble the rocks of the endocontact, and by aplites that constitute sharply falling as well as flat-lying dikes.

On the whole, in the composition of the Zhanchivlan pluton are included highly diverse rocks. Besides the three aforementioned types of granites (corresponding to the granite, alaskite, and subalkaline-leucogranite formations), here are established a more complete range of differentiations of lithium-fluoride granites of Sharakhadin type: alaskites, microcline-albitic, amazonite-albitic, and albite-lepidolitic granites, as well as various metasomatic formations, zwitters, microclinites, quartz-lepidolite greisens, albitites, quartz-muscovite greisens, tourmalinic rocks, and quartz veins with tin and tungsten mineralization. The first discoveries in Mongolia of tantalum-bearing and albite-lepidolite granites likewise occurred here.

CHAPTER 3

Associated Minerals and Geochemistry of Amazonite

3.1 ROCK-FORMING, SECONDARY, AND ACCESSORY MINERALS OF AMAZONITE-CONTAINING ROCKS

In the various genetic types of amazonite-containing rocks at present have been uncovered approximately 200 mineral species. Of this number, approximately 130 have been found only in one of the types of rocks; in the majority of those rocks, no more than 20 species occur consistently, and those found in all genetic types comprise only approximately 10 minerals. The greatest quantity of mineral forms (more than 70) is noted in pegmatite deposits of the alkaline-granite formation, and the least quantity (10–15) in pegmatites and metasomatites of the alaskite formation. The leading role in amazonite-containing rocks is played by silicates (approximately 40%), followed by oxides (20%), sulfides (8%), fluorides (8%), and phosphates (8%). In the amazonite-containing paragenesis, approximately 30 mineral forms have been established directly with amazonite (Table 3.1). The specific set of minerals in which a principal role is played by (besides silicates and oxides) sulfides, fluorides, and phosphates (Table 3.2), and likewise the particularities of their chemical composition (Table 3.3) are evidence of the duration and fairly late stage of formation of amazonite-containing paragenesis. These minerals usually contain Fe, Mn, rare earth elements (REE), Y, and Pb and are rich in volatile (OH, F, S, B) and in radioactive (Th, U, K, Rb) elements.

We will focus attention on a range of tendencies of the change in the contents and on the characteristics of certain typomorphic features of the minerals that occur in all genetic types of amazonite-containing rocks.

Potassium feldspar (of which amazonite is a variety) quantitatively predominates over other minerals in the majority of genetic types of amazonite-containing rocks. Its content commonly increases from the alaskite to the alkaline-granite formation. We note certain particularities of common potassium feldspars,

Table 3.1   Minerals Proven to Be Known in Direct Paragenesis with Amazonite

Mineral	Color	Frequency of occurrence
Microcline	White, gray, yellow, pink	High
Orthoclase	Gray	Low
Albite	White, bluish	High
Oligoclase	White, light-gray	Medium
Quartz	Smoky, gray, black	High
Biotite	Black	Medium
Muscovite	Light-gray, greenish, ruby	Low
Lithium micas (protolithionite, zinnwaldite, lepidolite)	Light-gray, silvery, lavender	High
Topaz	Colorless, bluish, yellow	High
Beryl	Bluish-green, greenish, yellowish	Medium
Phenakite	Colorless	Low
Genthelvite	Colorless, pinkish-lavender	Low
Gadolinite	Reddish-brownish	Low
Yttrialite	Black	Low
Tourmaline (schorl)	Black	High
Hydroxides of iron	Reddish-brownish	Medium
Magnetite	Black	Low
Hydroxides of manganese	Black	Low
Columbite-tantalite	Black	High
Pyrochlore-microlite	Red-brown, greenish	High
Cassiterite	Brown, red-brown	High
Fluorite	Purplish, green, yellow	High
Galena	Gray	High

Table 3.2   Characteristic Secondary (+) and Accessory (x) Minerals of Amazonite-Containing Rocks

Mineral	Mineral complexes of the rocks							
	1	2	3	4	5	6	7	8
I. Native elements								
Lead	–	–	–	x	x	–	x	–
Bismuth	–	–	–	x	x	–	x	–
II. Sulfides and related minerals								
Sphalerite	x	x	–	x	x	–	x	x
Galena	x	x	x	x	x	x	+	x
Molybdenite	x	–	–	–	x	–	x	x
Bismuthine	–	–	–	–	–	–	x	x
Pyrite	x	x	–	–	x	x	x	x
Arsenopyrite	x	x	–	–	x	x	x	x
Lillianite	–	–	x	–	–	–	–	x

Table 3.2   Characteristic Secondary (+) and Accessory (x) Minerals of Amazonite-Containing Rocks continued

Mineral	Mineral complexes of the rocks							
	1	2	3	4	5	6	7	8
III. Oxygen compounds								
A. Oxides and hydroxides								
Uraninite	x	−	−	−	−	−	x	−
Magnetite	x	x	−	−	−	x	x	x
Hematite	x	x	−	−	−	−	−	−
Hydroxides of iron	x	x	x	x	x	x	x	x
Ilmenite	x	x	−	x	x	x	x	x
Cassiterite	x	x	x	x	+	x	x	x
Rutile	x	x	−	x	−	−	x	x
Ilmenorutile-struverite	−	−	−	−	x	x	−	x
Columbite-tantalite	x	x	x	x	+	x	x	x
Pyrochlore-microlite	x	x	x	x	+	x	x	x
Fergusonite	x	−	−	−	x	−	x	x
Betafite	−	−	−	−	−	−	x	x
Samarskite	−	−	x	−	−	−	x	x
B. Oxysalts								
Silicates								
Almandine-spessartine	x	−	−	−	x	x	x	x
Zircon	x	x	x	x	x	x	x	x
Thorite	x	x	−	−	x	−	x	−
Titanite	x	−	−	−	−	x	x	−
Topaz	x	+	x	+	+	+	−	+
Orthite	x	−	−	−	−	−	x	x
Tscheffkinite	−	−	−	−	−	−	x	x
Beryl	x	x	x	−	x	x	−	x
Tourmaline	x	x	x	x	x	x	x	x
Aegirine	−	−	−	−	−	−	x	x
Hastingsite	−	−	−	−	−	−	x	x
Muscovite	x	x	−	x	x	−	−	−
Biotite	+	+	−	+	+	−	+	+
Lithium mica	−	x	+	+	+	+	x	x
Genthelvite	−	−	−	−	−	−	x	x
Gadolinite	−	−	−	−	−	−	x	x
Phosphates								
Xenotime	−	−	−	x	−	−	x	x
Monazite	x	x	−	x	x	−	x	x
Apatite	x	−	x	−	−	x	x	x
Carbonates								
Cerussite	−	−	−	−	−	−	x	x

Continued...

Table 3.2 **Characteristic Secondary (+) and Accessory (x) Minerals of Amazonite-Containing Rocks continued**

Mineral	Mineral complexes of the rocks							
	1	2	3	4	5	6	7	8
Sulfates								
Anglesite	–	–	–	–	–	–	x	x
Tungstates								
Wolframite	–	–	–	x	x	x	–	x
IV. Halides								
Fluorides								
Fluorite	x	x	x	x	x	x	+	x
Fluocerite	–	–	–	–	x	–	x	x

N.B.

1. Complexes of rocks of granitoid formations: (1–2) alaskite (1: pegmatite bodies in endo- and exocontacts of plutons, 2: local (endo- and exocontact) metasomatites and feldspar (amazonite)-quartz veins); (3–6) subalkaline-leucogranite (3: pegmatite veins not found to have a connection with granite plutons, 4–5: amazonitic granites and their metasomatites (4: amazonite-albite subtype, 5: albite-greisen subtype), 6: subvolcanic rocks (ongonites)); (7–8) alkaline-granite (7: rare-metal/rare-earth subtype, 8: rare-metal subtype).

2. The table does not include minerals occurring in any single (rarely two) genetic type of rocks. Among them are minerals that are common and fairly widely distributed in the rock, but not characteristic or atypical for amazonite-containing rocks, as well as minerals that are rare but highly indicative for certain amazonite-containing paragenesis (including new, comparatively recently discovered minerals, which are their own type of "endemics" of separate deposits of amazonite). These are comprised by the following minerals:

   I. **Native elements:** graphite, antimony, gold.

   II. **Sulfides and related minerals:** chalcosite, bornite, chalcopyrite, stannite, pyrrhotine, cosalite, goongarrite, löllingite

   III. **Oxygen compounds:**

      A. Oxides and hydroxides: thorianite, chrysoberyl, gahnite, bismuth, corund, pyrophanite, macedonite, loparite, davidite, murataite, anatase, mangantantalite, tapiolite, stibiotantalite, ixiolite, simpsonite, euxenite, aeschynite, priorite, risørite, yttrotantalite, tantite, natrobistantite, cesstibtantite, alumotantite, natrotantite, calciotantite.

      B. Oxysalts. Silicates: willemite, phenakite, cerphosphorhuttonite, hingganite, thalénite, keiviite, thortveitite, yttrialite, rowlandite, barylite, bavenite, bertrandite, leucophanite, astrophyllite-kupletskite, epidote, clinozoisite, spensite, kainosite, caysichite, vyuntspakhkite, aegirine-augite, bustamite, rhodonite, babingtonite, pyroxmangite, arfvedsonite, riebeckite, chlorite, kaolinite, dickite, montmorillonite, nontronite, hydromuscovite, orthoclase, petalite, pollucite, scapolite, holtite, helvine, danalite, apopyllite, laumontite, stilbite. Phosphates: pyromorphite, triplite, goyazite, amblygonite, lun'okite, triphylite, purpurite, delvauxite. Sulfates: baryte. Arsenates: scorodite. Vanadates: vanadinite, clinobisvanite. Borates: hambergite. Tungstates and molybdates: scheelite, wulfenite, chillagite.

   IV. **Halides.** Fluorides: sellaite, tveitite, cryolite, chiolite, prosopite, gearksutite, thomsenolite, pachnolite, cryolithionite, elpasolite, weberite, ralstonite.

directly according to which may develop amazonite color. Observations of variegated amazonites, as well as experiments according to the thermal decoloration of their uniformly colored varieties, is evidence that in the overwhelming majority of cases, amazonite tone is spatially and genetically associated with one of a range of feldspars tones of color intrinsic to potassium—from white or light-gray to pink or reddish-brownish—and likely does not generally manifest in connection with varieties of these minerals that have deeper and a more intensive color: red, brown, and black. Furthermore, according to the data of M.N. Ostrooumov, not infrequently, a definite connection of tones of color of amazonite and of

Table 3.3   Classes of Associated Minerals of Amazonite and Their Characteristic Chemical Elements

Classes of minerals	Number of mineral species	Characteristic elements
Native metals	5	Pb, Bi
Sulfides and related minerals	17	Pb, Fe, Bi, Zn
Oxides and hydroxides	44	Fe, Mn, Ti, Ta, Nb, Sn, U, Th, Pb, TR
Silicates	78	K, Na, Rb, Cs, Ti, Li, Fe, Mn, Be, Y, REE, Th, U, Pb, F, OH, S
Carbonates	9	Pb, Fe, Mn
Phosphates	11	U, Th, Pb, Tr (F, Cl, OH)
Tungstates	3	Fe, Mn, Sc
Arsenates	1	Fe, H_2O
Vanadates	2	Pb, Bi, Cl
Molybdates	1	Pb
Sulfates	2	Pb, Ba
Borates	1	Be, OH, F
Fluorides	16	Ca, Na, Al, OH, TR

common microcline is established. Thus, a saturated blue and bluish-green color of amazonite sections corresponds with a white and light-gray color of the non-amazonite sections. A green and yellowish-green color of amazonite often corresponds with yellow, reddish-brownish-yellow, and reddish-brownish microclines.

Considering the aforementioned, it may be proposed that with the presence of the other favorable conditions, the above-indicated intensive colors of microcline may be "prohibitive" for the manifestation of amazonite tone. However, it is evident that the "favorable" conditions for amazonitization of color ought, likewise, to be considered as necessary but insufficient on their own.

Albite is the most characteristic associated mineral in all types of rocks of the three granitoid formations. From the alaskite to the alkaline-granite formation, the role of albite increases. It is interesting to note that not a single manifestation of amazonite has been observed without notable albitization of the rocks, although albitization by itself is insufficient for the development of amazonitization. Albite associated with amazonite is commonly distinguished as milky-white, occasionally as bluish (the Sudetes, Pamirs, and elsewhere) color or iridescence (Kola Peninsula, Karelia, eastern Siberia—in these regions iridizes not albite, but albite-oligoclase).

Muscovite of "ruby" color and high industrial quality in significant quantities occurs in amazonite-containing pegmatites of India and Canada. Significantly more rarely, similar muscovite without industrial significance occurs in other types of pegmatites with amazonite. More common for alkaline-granite pegmatites is greenish muscovite in the form of fine crystals of low quality; while in veins with the brightest and most uniform amazonite, muscovite is practically absent. In granites and their connected metasomatites, muscovite is common for the formation of alaskites, replaces by lithium mica in the subalkaline-leucogranite formation, and is not characteristic for the alkaline-granite formation.

Lithium mica is a typical mineral of the subalkaline-leucogranite formation and a rare accessory in rocks of the other formations. In amazonitic granites, it may be represented by one or several of the members of a range of protolithionite-zinnwaldite-lepidolite. In sodium-lithium pegmatites with amazonite, lepidolite is observed most commonly.

Quartz. Amazonite not infrequently, particularly in pegmatites, is directly and likely genetically connected with quartz. Inner amazonite zones of crystals of microcline that border with blocky quartz are known, as well as amazonite "veinlets" with quartz that transverses microclines of common colors. In this connection, it is interesting to note a strengthening of the smoky tone of quartz, which is more distinctly manifest in connection with intensively amazonitized microclines (for example, the Pirtim deposit in Karelia). In the amazonites of the Kola Peninsula, M.N. Ostrooumov has observed veinlets or separate metacrystals of quartz superimposed on potassium feldspars with already manifested amazonite color, which distinctly manifests in the altering of the color of the amazonite.

In pegmatites of the alaskite formation, weakly colored amazonite can occur in rare cases in association with piezoquartz. Pegmatites with amazonite of the Indian type (see Table 2.2) contain predominantly smoky blocky quartz. Quartz of similar color is also commonly developed in amazonitic pegmatites of the alkaline-granite formation of the Kola type. In cavities of pegmatites with amazonite of the Ilmen-type, crystals of smoky quartz, morion, and occasionally amethyst are not rare.

Below, we consider the characteristic secondary and accessory minerals.

Beryl. Large crystals of beryl of yellowish-green color, more rarely of other colors, are noted in muscovite-bearing pegmatites with amazonite of India and Canada (the alaskite formation). Its polychromatic varieties are particular to the pegmatites of sodium-lithium type. This mineral is not characteristic for alkaline pegmatites of the Kola Peninsula, while in the pegmatites of the Il'menskie Mountains, amazonite it is not infrequently associated with aquamarine.

In granites and in their connected metasomatites, beryl, in rare cases, is observed in small quantities in connection with the subalkaline-leucogranite formation; it is not typical for the alkaline-granite formation and is episodic for amazonite manifestations of the alaskite formation.

Topaz practically does not occur in rare-metal/muscovite-bearing, sodium-lithium, and alkaline-granite pegmatites of the Kola type, that is, in all amazonite-containing rocks of the most ancient age and middle to large depths. In the relatively younger and less deep formations, the significance of topaz sharply grows: in miarolitic pegmatites of the alaskite formation are occasionally noted industrial concentrations of its gemstone varieties (but amazonite here has not yet been noted); topaz plays a highly significant role in amazonite granites and their genetic types (up to ongonites) of the subalkaline-leucogranite formation; finally, widely known are the large "heavyweight" crystals of various colors from the amazonitic pegmatites of the Il'menskie Mountains—precisely here, as has already been noted, miners have employed amazonite in the capacity of a prospecting indicator for topaz.

Fluorite is a highly typical mineral of amazonitic pegmatites of the Kola Peninsula (particularly yttrofluorite). It is observed more rarely in the pegmatites with amazonite of the Pikes Peak deposit (USA) and extremely rarely in the similar pegmatites of the Il'menskie Mountains. Thus, contrary tendencies are apparent in the altering of content of fluorite and topaz (the principle bearers of fluorine) in the aforementioned alkaline-granite pegmatites. In the subalkaline-leucogranite formation, fluorite is rare in pegmatites with pale amazonite belonging to the sodium-lithium type, and it is highly characteristic for all principal genetic types of small depths of this formation: amazonitic granites and their veined derivatives. In the alaskite formation, it occasionally forms industrial concentrations in miarolitic pegmatites (including optical varieties), but amazonite here already belongs to the rare minerals.

Cassiterite is a common accessory mineral of rocks of the subalkaline-leucogranite formation; moreover, in certain plutons of amazonitic granites of Cimmerian age, it has an industrial significance; in amazonite-containing rocks of the other formations, the quantity of cassiterite is markedly decreased.

Columbite-tantalite in rocks of the alaskite formation is represented principally by the niobium variety (Nb/Ta ratio from 5 to 13) and is present in significantly lesser quantities than in amazonite-containing rocks of the alkaline-granite formation, in which alongside tantalo-columbite (Nb/Ta ratio from 2 to 3, more rarely 1) occur minerals of this set that are even richer in tantalum. In the subalkaline-leucogranite formation are potentially industrial concentrations of columbite-tantalite and lower niobium-tantalum ratios.

Pyrochlore-microlite is the most typical mineral of amazonitic granites and their other generic types of the subalkaline-leucogranite formation. In it are noted the highest Nb/Ta ratios (from 1 to 5–6), and variations of the contents of Ta_2O_5 comprise 37–74%. In rocks of the alkaline-granite formation, this mineral is represented by lead-based varieties (contents of PbO reaches 40%). The mineral varieties of the given set, rich in niobium, are noted in the other genetic types of the alaskite formation.

Galena is the most characteristic sulfide of the amazonite-containing rocks of all three formations. Its content grows from the alaskite to the alkaline-granite formation, while the highest quantity has been recorded in the ancient pegmatites of the alkaline-granite formation. In certain of these (Broken Hill), galena can form fairly high concentrations.

Of the remaining accessory minerals, we will name only those particular to the defined formations and to types of amazonite-containing rocks. In the pegmatites of the alkaline formation are frequently developed such minerals as gadolinite, genthelvite, helvite, yttrialite, and talenite, and various tantalo-niobates, aluminofluorides, and bismuth sulfides, and sulfosalts. Moreover, for the Kola and Ilmen types defined by the author can also be noted a more narrow specialization. More characteristic for the Kola type are yttrium-containing (yttrialite, vyuntspakhkite, keiviite, and others) minerals, as well as gadolinite, genthelvite, hingganite; more frequently occurring in the Ilmen type are tantalo-niobates (columbite, pyrochlore-microlite), numerous aluminofluorides (cryolite, chiolite, weberite, and others), phenakite, and helvite; fergusonite and samarskite are intrinsic to those and to others, as well as to transitional types of pegmatites.

Only in certain pegmatites of sodium-lithium type with pale amazonite are noted spodumene, petalite, and amblygonite; on the whole, however, amazonite ought to be considered a highly uncharacteristic mineral for spodumene pegmatites.

Through the comparison of the mineral content of the principal genetic types of amazonite-containing rocks, it emerges that its variations in all types of pegmatites are determined principally by the quantitative correlations between the rock-forming minerals and to a lesser degree by secondary and rare-earth/rare-metal mineralization. Amazonite-containing granites are comprised in the majority of cases by the same minerals as compose pegmatites, but are distinguished from the latter by a more widespread development of lithium mica and the principally rare-metal paragenesis of accessory minerals (see Table 3.2).

3.2 GEOCHEMICAL PARTICULARITIES OF AMAZONITE-CONTAINING PARAGENESIS

Analyzing the geochemical particularities of amazonite-containing rocks, above all can be noted that the range of formations from alaskite to alkaline-granite corresponds with a regular change of the contents of certain petrogenic (major) elements: an increase in the contents of alkaline (with relative maximums of lithium, rubidium, and cesium in the subalkaline-leucogranite, and of sodium and potassium in the alkaline-granite formation), as well as of aluminum oxide, iron, and titanium, and with a certain decrease of silicon dioxide (Table 3.4, see Table 2.1), an increase of the agpaitic coefficient. Definite tendencies are revealed as well for rare and trace elements: from alaskite to alkaline-granite formations increase the contents of rare earths elements, yttrium, niobium and tantalum (the maximum for Ta is in the subalkaline-leucogranite, and for Nb – in the alkaline-granite formation), and fluorine. The occurrence of boron, tin, and other elements is more complicated: an elevation of their concentrations is noted toward the subalkaline-leucogranite formation, and subsequently a decrease toward the alkaline-granite (see Table 2.1).

Comparing granitoid rocks of monotype formations of variously aged series, it is not difficult to establish that in their later and increasingly young members

Table 3.4	Average Significances (X: Numerator) and Dispersion (Denominator) of Contents of Petrogenic (Major) Elements in Granitoids of Various Formation Types, wt%			
Oxides	Formation Type			
	Granite	Alaskite	Subalkaline leucogranite	Alkaline-granite
SiO_2	71.71 / 2.87	75.24 / 0.75	76.03 / 1.19	74.67 / 1.27
TiO_2	0.30 / 0.001	0.14 / 0.002	0.08 / 0.001	0.15 / 0.006
Al_2O_3	14.22 / 0.50	13.01 / 0.31	13.09 / 0.88	12.50 / 0.97
Fe_2O_3	1.00 / 0.26	0.75 / 0.12	0.43 / 0.10	1.31 / 0.36
FeO	1.73 / 0.40	0.95 / 0.18	0.93 / 0.16	1.09 / 0.21
MnO	0.06 / 0.0005	0.04 / 0.0003	0.03 / 0.0002	0.04 / 0.0003
MgO	0.68 / 0.09	0.26 / 0.02	0.15 / 0.01	0.20 / 0.02
CaO	1.62 / 0.36	0.74 / 0.08	0.45 / 0.06	0.51 / 0.08
Na_2O	3.59 / 0.14	3.68 / 0.13	4.30 / 0.43	4.50 / 0.24
K_2O	4.20 / 0.19	4.63 / 0.13	4.71 / 0.15	4.57 / 0.19
Number of objects (of petrochemical selections)	192	215	55	103

After Ref. [4].

occurs a concentration of fluorine and a range of lithophylic rare elements (F, Li, Rb, Ta, Nb, Tl, Hf, Pb—particularly the first five) and a decrease in the concentration of barium, strontium, rare-earth elements, and zirconium. This same tendency is traced with the transition from the beginning phases of various complexes and formations toward the final and from the early generations of rock-forming, secondary, and accessory minerals to the later. Consequently, in the granitoid rocks of all rare-metal-bearing formations from early to late members of evolutionary ranges, in the general case is noted an increase in contents of some, commonly low Clark (abundance ratio), elements, and a decrease in concentrations of others of relatively high Clark elements.

We will consider in greater detail the geochemical features of the principal genetic types of amazonite-containing rocks: granites and pegmatites. In the very first works dealing with amazonitic granites, they were classified to the group of specific plumasitic leucocratic granites, commonly high in aluminum oxide, with a significant quantity of albite and lithium mica, not having analogues among the traditional petrochemical types of granites, according to R. Daly. Therefore, it has been proposed to name them subalkaline leucogranites or "subalkaline" granites [4].

In the structurally complex amazonite plutons, granites of various zones markedly differ in terms of chemical composition (according to content of petrogenic major elements as well as of element-impurities), as well as the quantitative and qualitative (chemical) content of rock-forming and accessory minerals.

According to the data of A.A. Beus, L.G. Fel'dman, and other researchers [2,47], biotite granites constituting deep horizons of amazonite-containing plutons in the petrochemical regard represent rocks of subalkaline range, which are characterized by very low contents of iron, calcium, magnesium, titanium, and phosphorus, and high contents of fluorine. The Clark contents of rare elements are notably higher in them than in the all genetic types of granite formations: tantalum, rubidium, and lead by 2 to 3 times, niobium, and lithium by 4 to 5 times, tin by 7 times, etc. Rare elements in these rocks are concentrated principally in columbite, mica, and cassiterite.

With the transition from biotite granites toward the amazonitic facies of the upper horizons of the plutons occurs a general elevation of the alkalinity of the rocks (with the leading role of sodium), an increase in contents of aluminum oxide, fluorine, lithium, rubidium, tantalum, and tin, and a decrease of silicon oxide, iron, calcium, magnesium, barium, strontium, zirconium, and rare-earth elements. Moreover, there is also a decrease in the significance of indicatory ratios (K/Rb, Nb/Ta, Zr/Hf), a change in the chemical composition of mica (from biotite to micas of the lithium range); in the latter are accumulated fluorine and lithium, and the concentrations of tantalum and niobium decrease.

With the replacement of earlier rocks of the deep horizons by later rocks (by measure of proximity to the surface), there is a decrease in the quantity (up to disappearance) of titanium- and cerium-containing minerals: ilmenite, anatase, and monazite, and an increase in the contents of topaz, fluorite, tantalo-niobates, and sulfides (galena, sphalerite, and bismuthine). In rare cases, there have been finds of native lead and bismuth, aluminofluorides, uranitite, bastnesite, fluorocerite, and others. In accessory minerals of the peripheral parts has been established an elevated content of tantalum, niobium, and tin in relation to analogous minerals from the rocks of deeper horizons. In the latest rocks, 80% of the tantalum is concentrated in tantalo-niobates, the maximum content of lithium is noted in mica, and of rubidium in mica and in amazonite. As a result, in the endocontact and peripheral sections of the plutons of the subalkaline-leucogranite formation occurs a concentration of a range of rare elements; their contents significantly exceed the Clark abundance ratio: Ta by 15 to 20 times, Nb and Sn by 5 to 10 times, Li by 10 to 15 times, Rb by 6 times, and Tl by 5 times.

Ongonite rocks are close in terms of petrochemical particularities to amazonite-albite and albite-lepidolite rare-metal granites. Amazonite-containing ongonites are characterized by analogous amazonitic granites, concentrations of rare and ore elements, elevated fluorine, lithium, rubidium, tantalum, niobium, beryllium, hafnium, and lead, and decreased barium, strontium, rare-earth elements, and zirconium.

Amazonitic pegmatites of various formations are as a rule quite close in terms of geochemical particularities to the described genetic types of rocks of the subalkaline-leucogranite formation. This has been established particularly distinctly for pegmatites of the alkaline-granite formation, where the leading rare elements are fluorine, lithium, rubidium, tantalum, and niobium. Furthermore, in these pegmatites, a definite development occurs for rare-earth and beryllium mineralization, not typical for amazonitic granites and ongonites. Characteristically, in the ancient and deepest amazonitic pegmatites, rare-earth elements and yttrium predominate over tantalum, niobium, and beryllium. An inverse correlation between REE and rare-metal mineralization has been established in the younger and less deep amazonitic pegmatites. Moreover, certain subadjacent elements, in part fluorine and beryllium, form various mineral species. Thus, in late Proterozoic amazonitic pegmatites, minerals of these elements are represented principally by fluorite, gadolinite, genthelvite, and hingganite, and in the late Paleozoic by topaz, various aluminofluorides, and beryl.

A certain geochemical direction in the course of the pegmatite process is expressed by a change of the composition of minerals and their associations over time. A general particularity of the process of formation of amazonitic pegmatites (from the peripheral zones of the vein bodies toward the inner zones) is an increase in content of rare alkalines and a decrease in barium and strontium in feldspars, a concentration of fluorine and lithium in micas, and a growth of concentrations of tantalum, tin, lead, manganese, and yttrium in accessory minerals.

For amazonitic pegmatites of the alkaline-granite formation, beyond the dependence on their age and depth have been established analogous regularities of development of rare-metal mineralization. In the first order, this concerns tantalo-niobates (columbite-tantalite, pyrochlore-microlite, samarskite, fergusonite, and others). With the transition from early generations of these minerals toward the late generations, there is a marked increase in the role of tantalum, tin, lead, uranium, and thorium in their composition, up to the manifestation in the latest stages of specific tin (Pb-microlite) and tantalum (Ta-samarskite) varieties. Additionally, the composition of these minerals includes elements not entirely common for pegmatites—bismuth, tin, and tungsten—which likewise commonly form particular mineral species (bismuth, bismuth sulfosalts, cassiterite, and wolframite). It must be mentioned that bismuth and tin-zinc (galena, sphalerite) sulfide mineralization is fairly typical for amazonite-containing pegmatites of the alkaline-granite formation.

In the final steps of the post-magmatic stage of mineralogenesis in amazonitic pegmatites, there is also an increase in the role of yttrium, phosphorus, and manganese. This is expressed by an increase of the concentration of yttrium in all rare-earth minerals up to the inception of its particular mineral forms (kuliokite, vyuntspakhkite, keiviite, and others) and in the distribution of tantalo-niobates and garnets with high content of manganese, and subsequently of various phosphates (triplite, xenotime, and others).

From all of the above, it follows that the petrochemical evolution of the composition of the mineral-forming environment in the formation process of various genetic types of amazonite-containing rocks has been strikingly close and has been characterized on the whole by unidirectional and unambiguous tendencies in the change of compositions of rock-forming, secondary, and accessory minerals. Fairly close as well are the petrochemical and geochemical particularities of amazonitic granites and pegmatites. Particular to the former as much as to the latter is an elevated content of alkaline, silicon oxide, and aluminum oxide, and also of iron in pegmatites. No less indicative a factor in providing evidence of the great similarity of these rocks is the establishment in them of high concentrations and identical composition of rare elements. Their typical rare alkaline elements and fluorine accumulate markedly in amazonite-containing (late) paragenesis of granite and pegmatite rocks. The basic mass of these elements (apart from the fluorine forming its own minerals) is concentrated in micas; moreover, micas in amazonitic granites and pegmatites are similar in terms of lithium, rubidium, cesium, and fluorine content.

The distribution of fluorine cannot be left unmentioned. The principal and consistent mineral-bearers of fluorine in amazonite-containing rocks are fluorite and topaz. As mentioned above, these different mineral species of fluorine to a certain degree are isolated in amazonitic granites and pegmatites that vary in terms of their depth and age of formation. For example, in the more ancient (Precambrian) and deep pegmatites (Kola Peninsula, eastern Siberia), fluorine, partially dispersed in micas, occurs exclusively in the form of fluorite, whereas in relatively small depth and younger (Paleozoic) pegmatites (southern Ural, Russia; Pikes Peak, USA), its leading mineral form is represented by topaz. The same tendency is noted in amazonitic granites: in the Cimmerian granites

(Transbaikal, Mongolia), topaz serves as the principal mineral-bearer of fluorine, while Caledonian granites (Kyrgyzstan) along with topaz contain a significant quantity of fluorite.

And finally, in the distribution of accessory rare elements, many common features can be found in amazonitic granites and pegmatites. In both occur niobium, tantalum, tin, yttrium, rare-earth elements, beryllium, and thorium, which either form particular mineral species, or are found in a dispersed state. It must, however, be kept in mind that among the noted elements, some, for example, rare-earth elements and beryllium, are more characteristic for Precambrian alkaline-granite pegmatites (Kola Peninsula); whereas others, in part tantalum, niobium, and tin, are characteristic for Mesozoic granites of the subalkaline-leucogranite formation. Of course, this does not imply the total absence of tantalo-niobates and cassiterite in the noted pegmatites, although those minerals have subordinate significance in relation to yttrium-beryllium rare-earth minerals; furthermore, tantalo-niobates (columbite-tantalite, pyrochlore-microlite) and other accessory minerals (cassiterite, fluorite, fluorocerite, thorite, and monazite) contain elevated quantities of yttrium and heavy lanthanides.

It is indicative that in certain Paleozoic amazonite-containing granites (Kazakhstan, Kyrgyzstan) and pegmatites (Urals; Pikes Peak, USA) are noted rare-earth (samarskite) as well as rare-metal (columbite) minerals. Attracting attention in granites and pegmatites with amazonite are frequent finds of accessory minerals of radioactive elements: thorite, uranothorite, and occasionally uraninite; impurities of uranium and thorium are highly characteristic as well for many rare-earth and rare-metal minerals of these rocks.

CHAPTER 4

Morphology, Crystal Chemistry, and Properties of Amazonite

4.1 MORPHOLOGY AND ANATOMY

Amazonite in the form of more or less regularly facetted crystals occurs principally in the cavities of pegmatite bodies, in fine blebs of endo- and exocontact zones of granite plutons, as well as among albitites and within ongonites [18]. The crystals are commonly weakly extended along [001] and sometimes markedly tabular habit parallel to {010}; more rarely, an isometric habit is observed. The dimensions of the crystal along the greatest measurement fluctuate from several millimeters to 30–40 cm. It is interesting to note that the largest crystals of amazonite have been found in the most ancient Proterozoic pegmatites (Kola Peninsula), whereas principally characteristic for the young Cretaceous formations are minimally sized crystalline individuals (Transbaikal, Mongolia).

Data from various publications and author's investigations [18,21] indicate significant similarities in the habitus of crystals of amazonite from widely different deposits (Fig. 4.1). The selection of simple forms in crystals from pegmatites of western Keivy, the Il'menskie Mountains, the Pikes Peak area, and the quartz-amazonite veins of Transbaikal on the whole is analogous and, in order of decreasing area of the faces, commonly includes the following: m {110}, b {010}, c {001}, x {101}, z {130}, o {111}, y {201}, and n {021} (Fig. 4.1). Simple forms have been reported here for the monoclinic system, insofar as the symmetry of the sculpture of the faces (growth and vicinal forms) corresponds to the monoclinic crystal class symmetry (L$_2$PC). Paramorphs of microcline are

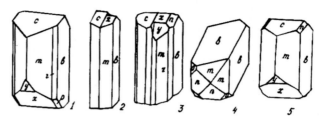

FIGURE 4.1
Forms of crystals and twins of amazonite. Crystals from Il'menskie Mountains (1), Transbaikal (2), Pikes Peak (3), Baveno (4), and Carlsbad (5) twins.

observed along the orthoclase that manifests in the optical picture of the cross-hatched twinning—twins appearing during the triclinization of orthoclase.

Found in accordance with the habitus of crystals of amazonite is a syngenetic sculpture of their faces. Thus, according to the data of V.A. Popov and M.N. Ostrooumov [21], the crystals extended along the vertical axis, which are formed as a result of the primary growth of the faces c {001} and x {101}, have on these faces either a multitude of coarse isometric growth forms or a sculpture of multiple-headed surfaces. On the faces of the vertical belt, growth forms of such crystals are strongly extended in a vertical direction. Only in the very latest (low-temperature) crystals of amazonite on the faces c {001} and x {101} occasionally are observed growth forms not having planar symmetry—potentially evidence of the transition to crystallization of the triclinic phase of potassium feldspar. It can be proposed that triclinic symmetry is already intrinsic to part of the "structural particles" in the environment of crystallization.

The most frequent growth twinnings in amazonite are Carlsbad and Baveno. Occurring together with amazonite, albite of perthites and cleavelandite are twinned according to Carlsbad and albite laws, with other twinnings occurring more rarely. Twins of amazonite according to the Carlsbad law are commonly extended parallel to the axis [001] and according to the Baveno law parallel to axis [100]. The cross-hatched twinning typical for amazonite (microcline twinning) will be further described below.

Finds of amazonitic crystals are exotic; in nature, other forms of crystallization of this mineral predominate. Most characteristic for pegmatite veins are blocky deposits of potassium feldspar with amazonitic color, the dimensions of which occasionally reach several meters; their aspect is commonly close to isometric. Monomineral individuals of such amazonite are concentrated as a rule in the central (axial) parts of the pegmatite bodies, where they display maximum dimensions and their greatest (apart from cavity crystals) intensity of color. For individuals of blocky microcline-amazonite, particular faces are observed only from the side of blocky quartz (commonly in the axial part of the vein) or in separate narrow zones of lixiviation and subsequent regeneration. In the intermediate and peripheral zones of pegmatite bodies with the graphic texture particular to them, amazonitic color of potassium feldspar (if it is observed at all) markedly recedes in intensity as much as in the volume of distribution of blocky varieties from the central zone. Here amazonitic color is commonly manifested in the form of stains in contrast to some common color of potassium feldspar, the boundaries of which are highly variable and conventional (gradual). In cases with weakly developed amazonitic coloration, similar individuals of potassium feldspar generally seem non-amazonitic to the untrained eye.

In granitic plutons, besides fine, regularly faced, small crystals from the cavities of lixiviation, are noted two to three additional morphological types of crystals of amazonite: crystals idiomorphic relative to quartz, those extending from particularly amazonitic granites (of the order of several millimeters in size), and distinctly elongated (length from hundredths of a centimeter to 10–20 cm) porphyritic crystals among fine-grained aplitic rock. The latter are distinguished by a nearly square or right-angled cross-section.

In subvolcanic rocks (ongonites) amazonite constitutes porphyritic grains with dimension of 0.7 mm in diameter among a fine- or pseudo-crystalline matrix. They are characterized by isometric (more rarely tabular or elongated) form, commonly having distinct and regular right-angle lineaments occasionally in the form of a truncated rhombus.

On the whole, as indicated by the data in the literature and by the observations of the author, amazonite in its morphological aspects does not possess any specific characteristics distinct from microcline and orthoclase of common (non-amazonitic) color. Here, it is also appropriate to note that the character of the inductive surface of the junction of individuals of potassium feldspar-amazonite with some mineral (e.g., albite, quartz, or topaz) "in itself," that is, without involvement of additional data, does not enable unambiguous judgment of whether another mineral has synchronously grown with some variety of potassium feldspar (commonly colored or amazonitic). With rare exception (e.g., mineral individuals of amazonite from one of the pegmatite veins of western Keivy), a common particularity of amazonitic feldspar is its variegation: a spotty and commonly irregular manifestation of amazonitic color within the boundaries of a separate mineral grain or crystal.

The phenomenon of the growth, change of form, composition, and properties, as well as dissolution and regeneration are commonly reported in the anatomical picture of mineral individuals. In this connection, the author researched amazonitic grains and crystals from various deposits in horizontal crystallographically oriented cross-sections, prepared samples of hundreds of mineral-incrustations in amazonites, and obtained neutron-activation radiography. Particular attention was devoted to the study of the ontogenic particularities of amazonite with the aim of an interpretation of its formation history. Principal observations were performed on rock material from the pegmatites of Ploskaia Mountain (western Keivy), the Il'menskie Mountains, and the granites of Transbaikal. Insofar as the most essential aspects of the genesis of amazonite are connected with the appearance and evolution of their color and perthitic growths of albite, the study of these particularities is imparted with the greatest significance.

Already over the course of field observations of amazonites of one of the veins of Ploskaia Mountain, a wide variation of colors and perthitic intergrowth of

PHOTO 7
Variability of form, size, and quantity of ingrowths of albite in variously colored amazonites (Kola Peninsula, western Keivy).

amazonites has been noted [11]. Specialized research has enabled the discovery of a definite regularity to their combination, revealing that in a series of variously colored amazonites with a transition from blue to green tone, there is a corresponding gradual decrease in quantity of albite in ingrowths, a change of the crystallographic orientation, and a decrease in size of the latter, as well as of their isometrization (Photo 7). In terms of perthitic intergrowth, the studied amazonites can be divided into the four following types.

Amazonites of the first type are distinguished by a maximum saturation of perthitic intergrowth of albite: from 40 to 45 wt% of the area of a thin section (Photo 8). The ingrowths of platy form and large dimensions—thickness of 1–2 mm, length of 15–20 mm, more rarely greater—are distributed evenly and very closely to one another at a distance no greater than 4–6 mm. On the plane {001} of the ingrowths, a reticulate pattern forms.

Characteristic for amazonites of the second type are several ingrowths larger in comparison with the above-described—their thickness is from 1 to 4 mm, length from 5 to 35 mm, occasionally greater; however, they are situated in isolation

(a)   **(b)**

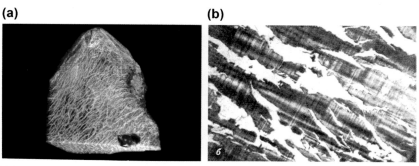

PHOTO 8
Early blue amazonite with platy lamellar ingrowths of albite: (a) polished hand sample, (b) thin section, magnification 24×, with analyzer.

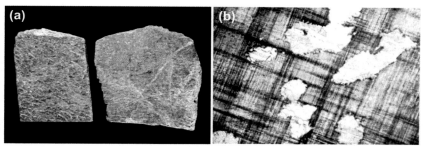

PHOTO 9
Late green amazonite with fine-stained isometric form albitic ingrowths: (a) polished hand sample, (b) thin section, magnification 24x, with analyzer.

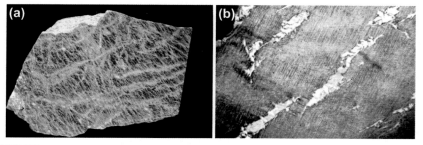

PHOTO 10
Emerald-green amazonite with fine-lamellar ingrowths of albite of filamentous habitus: (a) polished hand sample, (b) thin section, magnification 24x, with analyzer.

from one another in subparallel orientation. Such ingrowths occupy from 22 to 30 wt% of the area of the thin section (Photo 11).

Amazonites of the third type contain fine-stained ingrowths of albite of isometric or subisometric form in size, in the greatest measure from tenths of a millimeter to 1–2 mm. The quantity of the ingrowths is 10–15 wt% (Photo 9).

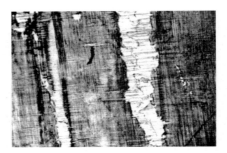

PHOTO 11
Lamellar subparallel albitic ingrowths (light) in amazonite. Thin section, magnification 24x, with analyzer.

Amazonites of the fourth type have the least quantity of albite in ingrowths (5–12 wt%), which form fine filamentous (in the plane of the thin section) or subisometric (point) microperthites, reaching in plane {001} several millimeters in length and tenths of a millimeter in width (Photo 10).

In the first of the enumerated types, ingrowths of albite are oriented parallel to {110}, and in the second are oriented approximately along {100}; in the third and fourth types, due to the subisometric habit, their orientation is expressed weakly, but on the whole is analogous to that of the second type. Between the identified types of perthites in amazonites are noted gradual transitions, occasionally established within the limits of a single grain or crystal. Such transitions are most characteristic for amazonites of the first three types. In large blocky individuals of amazonite in natural rock exposure can occasionally be noted a closely connected, parallel, and highly gradual change of perthitic ingrowths from the first to the third type and of color from blue to bluish-green; similar changes are traced within the boundaries of separate mineral individuals. It is indicative that the direction of such evolution of amazonites proceeds from the upper part of the vein toward its lower zones.

A regular and complementary change of color and perthitic structure of amazonites has been established reliably in a fairly representative series of samples. In separate small (on the order of 10–20 cm) samples of amazonite not infrequently can be seen the combination of two or three neighboring types of perthitic ingrowths (with one type predominating), but with a single tone of color (Photos 12 and 22). This suggests there is not a firm relation between color and perthites in the described amazonites. Furthermore, insofar as blue colors are observed in amazonites with perthites of the first type while green colors are observed in the perthites of the fourth type, cases of the inverse combinations are unknown; we can speak with certainty of a correlation between the two most important characteristics of amazonites.

Along with the gradual transitions between the described types of perthite-amazonites, not infrequently occurring in veins are interesting "transverses" of their correlation. These are formations having the form of veinlets with sharp borders (Photos 13 and 14), in which are commonly noted perthites of the third or fourth

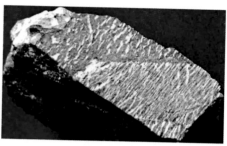

PHOTO 12
Zonal distribution of ingrowths of albite in blue early amazonite. Cleavage fragment along {001} of a crystal of amazonite; zones of growth along {100} and {130}.

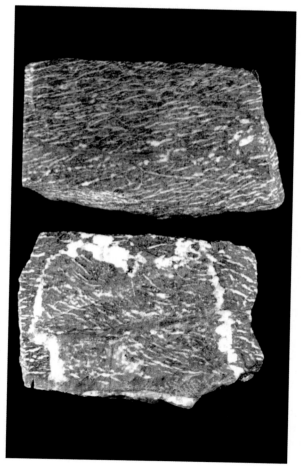

PHOTO 13
Cross-cutting interrelations of early and late segregations of amazonite of various color.

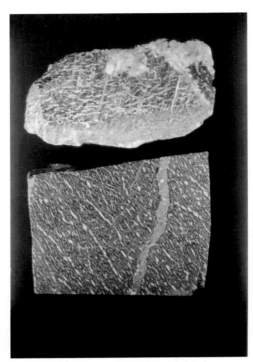

PHOTO 14
Development of the latest segregations of amazonite of yellowish-green and greenish-yellow colors along the early segregations of this mineral.

type in amazonite with green tones of color, whereas the amazonite enclosing them is represented by blue or greenish-blue varieties of the first (and more frequently of the second) type. The thickness of similar veinlets varies from several millimeters to several centimeters; their extension as a rule reaches no less than several or a few dozen centimeters. A highly characteristic particularity of the "veinlets" is a completely unbroken, monocrystallic structure of amazonite material within them and beyond their boundaries, which is easily established by a single reflection of the cleavage (see Photo 14). The difference is manifested only in color and perthitic ingrowths, which commonly appears as highly contrasting along the various sides of the contact of the veinlet. Inside similar formations are occasionally observed separate small skeletal crystals of gray or dark-gray quartz (Fig. 4.2).

The presence of veinlet forms of amazonite and of certain zoned grains and crystals (Fig. 4.3) leaves no doubts as to the relative age of its varieties of various color and varying perthitic structure: green colors and subisometric fine ingrowths of perthitic albite on the whole are intrinsic to generations of potassium feldspar later by comparison with varieties of this mineral with blue colors and its associated large lamellar ingrowths of albite (we are not yet discussing the question of the mode of occurrence and relative time of appearance of color and perthites).

blue amazonitic color. Consequently, in the Il'menskie Mountains have been noted not two, but one (blue) stage of amazonite-formation, which evidently can be compared with the early stage of amazonitization in the western Keivy pegmatite district.

We will indicate still one particularity of Il'menskie amazonites: their cross-hatched twinning (microcline twinning). In distinction from Keivy amazonites, in which twinning can be detected only under the microscope and almost exclusively in blocky varieties, in the Il'menskie amazonites, consequent to the formation of fine pores, it cannot infrequently be seen macroscopically on plane [001] and, evidently, in all three morphological types. In the initial stage of the manifestation of amazonitic color (the first and second types), the cross-hatched twinning renders a controlling influence on the distribution of amazonitic color—the latter develops, as was first noted in 1943 by A.N. Zavaritsky, principally in one crystalline individual of a twinned growth. Important also, as noted by A.N. Zavaritsky, is the coincidence of the pattern of cross-hatched twinning (as well as of cleavage) in sections with amazonitic color and without them. Finally, according to the observations of the author of the present research [21], the degree of the macroscopic manifestation of cross-hatched twinning is commonly connected to an inverse correlation with the intensity of amazonitic color.

The inverse correlation of the intensity of amazonitic color with the size and quantity of perthitic ingrowths of albite already has been noted, and it is established through the comparison of the three principal morphological types of amazonites. To this, it is necessary to add that in amazonite of the first type (with the most weakly manifested color), in separate cases, large perthitic ingrowths control the distribution of amazonitic color (as observed by A.N. Zavaritsky).

In conclusion, we briefly consider certain principal particularities of the inner structure (anatomy) of amazonites in granites. Here, as has already been noted, can be separated the following types of amazonite: grains and crystals in the matrix of granite (as a rock-forming mineral), porphyritic crystals in albitite, and fine small crystals in the cavities of lixiviation of pegmatoid zones and vein bodies. Particular to amazonites of the two latter types are a commonly zonal and sectorial structure of mineral individuals manifested in an irregularity and alternation of colors: blue or greenish amazonitic and light-gray or yellowish common color (Fig. 4.5, Photos 17 and 18). The sequence of the replacement of zones is most different in cavities where the peripheral zones of the crystals are most frequently non-amazonitic.

The distinct rectilinear pattern of zonality is occasionally complicated by means of stains of complex form with bluish and white color that are irregularly distributed in the crystals. Zones of growth are occasionally accentuated by grains of albite, quartz, and zinnwaldite. In the sectorial colored crystals of amazonite, pyramids of intumescence $\langle 110 \rangle$, $\langle 001 \rangle$, and $\langle 101 \rangle$ have greenish-blue or bluish-green color, $\langle 010 \rangle$ and $\langle 130 \rangle$ are colorless or white, and $\langle 201 \rangle$ is colored zonally. Occasionally, pyramids of intumescence $\langle 010 \rangle$ are frayed or in blocks.

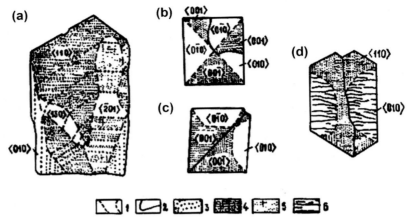

FIGURE 4.5
Zonality and sectoriality of color in crystals and twins of amazonite (Western pluton, eastern Siberia).
(a–d) Various sections: a, d—along (001); b, c—along (100). 1—Borders of pyramids of growth of planes;
2—twinning joints (twins: b, c—Baveno, d—Carlsbad); 3—growths of albite along the zones of growth;
4, 5—amazonite (4—greenish-blue, 5—pale-colored); 6—colorless microcline.

PHOTO 17
Zonal and sectorial monocrystal of amazonite.

PHOTO 18
Zonal-sectorial individuals of amazonite in fine-grained aplitotype veins (Western pluton, eastern Siberia).

Many crystals of amazonite with stained color that at first glance seem to be distributed nonuniformly turn out to be twins according to the Carlsbad, Manebach, or Baveno law. The latest generation of amazonite of bluish color (in paragenesis with smoky quartz and mica) most frequently is non-zonal and non-sectorial (Photo 19). A similar stained color of amazonite hinders the visual definition of its quantity in granites.

PHOTO 19
Blue amazonite of the late generation (non-zonal monocrystal from a quartz-amazonite vein of the endo-contact of the Western pluton, eastern Siberia).

In the crystals of rock-forming amazonite, by measure of strengthening intensity of amazonitic color (from the early to the late generations), according to the observations of the author and the data in the literature, is detected a progressively distinct cross-hatched twinning. In microcline occurs a "clearing away" of the primary ingrowths of albite, which probably is resettled in the form of lamellar crystalline grains inside those same grains of microcline or near to them. On the whole, the intensity of amazonitic color grows not only in granites from the early to the late grains and crystals, but likewise from the granitic to the pegmatoid (stockscheiders) and quartz-amazonitic (hydrothermal) forms of potassium feldspar (Photo 20).

PHOTO 20
Maximally intensively colored crystals of amazonite (quartz-amazonite vein, Western pluton).

4.2   CHEMICAL COMPOSITION

Despite the great number of research treating the chemical composition of amazonite, to date, limits have not been established for the isomorphic capacity of this mineral in ratio to rare and trace elements and their characteristic concentrations in amazonite from the rocks of various genetic and formational types. In the present section is analyzed all known data on the chemical composition of amazonites from the literature, additionally taking into account 42 full wet chemical, 157 partial X-ray fluorescence, and more than 500 optical spectrographic (of them half are quantitative) analyses of amazonites and the associated potassium feldspars obtained in the course of specialized research.

The results of the chemical analyses of amazonites provide evidence of the variability and mutability of the composition of this variety of alkaline feldspars. The chemical composition of amazonite changes within wide limits, but displays a distinct compositional evolution (Tables 4.1–4.5). Moreover, alongside the wide limits of the contents of impurity elements is also observed a certain variation of the ratio between main major elements. Attentive consideration of the analytical results provides convincing evidence first and foremost for the following [13]:

1. Notable variations in the composition of amazonites are noted only for two major elements: potassium and sodium. This is manifest with particular distinction in pegmatites of the alkaline-granite formation. According to the data of chemical analysis (Table 4.1), the content of K_2O in amazonites of the Kola Peninsula changes from 10.8 to 14.6 wt%, and the content of Na_2O from 1.53 to 4.2 wt%. The range of the main components was 63–87 mol% orthoclase (Or) and 13–37 mol% albite (Ab).

   Highly indicative is the inverse ratio established in amazonites between concentrations of the analyzed oxides. Such mutually exclusive behavior of potassium and sodium in all types of pegmatites (in terms of thickness, a fall depending on the distance from the magma chamber) has already long been known and many times noted by researchers (A.E. Fersman, N.A. Solodov et al.). Amazonitic pegmatites are likewise not exempted from this rule. Analysis of the samples selected in terms of thickness (from the peripheral to the central zones) in one of the pegmatite bodies (vein 19) of a section of Ploskaia Mountain has indicated that a distinct change in the composition of amazonites (Table 4.1) was connected primarily with the gradual decrease in content and dimensions of albitic ingrowths. A minimal content of albitic components was determined in grains and crystals of amazonite from the central zones and cavity of this pegmatite vein.

   A similar occurrence, which can be called the deperthitization of potassium feldspars, has been established in many amazonite-containing pegmatites that developed in the Kola Peninsula. Distinctly correlated with them is a change of the color of amazonites (see Section 4.1). Deperthitization corresponds to a lowering (deepening) of the color tone of amazonitic

Table 4.1 Chemical Composition of Amazonites from Alkaline–Granite Pegmatites of the Kola Peninsula, wt%

Textural zone of pegmatite	Color	No.	λ^a (nm)	SiO_2	Al_2O_3	Fe_2O_3	K_2O	Na_2O	Rb_2O	Cs_2O	Pb	K/Na	K/Rb	K/Cs	K/Pb	Rb/Cs
Pegmatoid	Blue	241	497	66.20	18.9	0.06	10.8	4.20	0.23	0.028	0.082	2.57	47	386	109	8.2
Pegmatoid	Blue	113	495	66.06	18.7	0.07	11.8	3.68	0.23	0.027	0.041	3.23	51	437	239	8.5
Pegmatoid	Greenish-blue	240	501	65.29	18.8	0.09	13.5	2.28	0.34	0.022	0.10	5.93	40	614	112	15.4
Pegmatoid	Greenish-blue	242	500	65.53	18.8	0.08	13.2	2.56	0.27	0.012	0.11	5.15	40	1110	100	22.4
Fine-blocky	Grass-green	249	512	64.97	18.5	0.03	14.0	2.08	0.44	0.019	0.30	6.73	32	736	39	23.2
Fine-blocky	Grass-green	292	514	65.32	18.6	0.04	14.0	1.72	0.51	0.027	0.36	8.14	27	518	32	18.9
Fine-blocky	Grass-green	103	542	65.0	18.8	0.03	14.2	1.53	0.65	0.036	0.73	8.99	22	395	16	18

[a]λ = color tone, quantitative parameter of color in the XYZ color system.

Table 4.2 Chemical Composition of Microclines and Amazonites from Various Pegmatite Deposits, wt%

Location, vein	Textural zone of pegmatite, number of sample	Color	λ[a] (nm)	SiO$_2$	Al$_2$O$_3$	Fe$_2$O$_3$	K$_2$O	Na$_2$O	Rb$_2$O	Cs$_2$O	Pb	K/Na	K/Rb	K/Cs	K/Pb	Rb/Cs
Kola Peninsula, Ploskaia Mountain, 20	Pegmatoid, 355	Pink	596	65.90	19.54	0.19	11.10	3.00	0.10	0.001	0.0004	3.08	111	1111	2303	100
Kola Peninsula, Ploskaia Mountain, 20	Pegmatoid, 345	Bluish-pink	500	66.10	19.10	0.24	11.70	3.20	0.16	0.002	0.02	3.56	71	5700	486	80
Kola Peninsula, Ploskaia Mountain, 20	Pegmatoid, 345	Blue	497	65.53	18.80	0.08	13.20	2.56	0.27	0.012	0.05	5.15	49	1100	219	23
Karelia, Pirtima, vein 1	Blocky, 172	Cream	590	64.40	18.65	0.08	13.14	3.05	0.07	0.002	0.01	4.3	188	6570	1065	35
Karelia, Pirtima, vein 1	Blocky, 163	Pale-green	568	65.43	18.6	0.08	13.8	2.32	0.20	0.001	0.02	5.95	69	1255	573	182
Karelia, Pirtima, vein 1	Blocky, 164	Bright-green	544	63.8	19.0	0.08	14.57	1.70	0.36	0.008	0.03	8.56	41	1890	366	46
Il'menskie Mountains, mine 70	Medium-grained, 70-84-1	Yellowish	593	–	–	–	10.15	4.56	0.087	0.004	0.007	2.23	117	2540	1203	22
Il'menskie Mountains, mine 70	Pegmatoid, 70-84-3	Yellowish-green	569	–	–	–	10.75	3.32	0.29	0.005	0.03	3.24	37	216	297	58
Il'menskie Mountains, mein 77	Medium-grained, 77a	Grayish-pink	587	–	–	–	11.75	4.87	0.11	0.006	0.045	2.41	107	1960	650	18
Il'menskie Mountains, mein 77	Graphic, 77-2	Pale-green	570	–	–	–	10.5	4.62	0.14	0.006	0.03	2.28	75	1750	291	23
Il'menskie Mountains, mein 77	Pegmatoid, 77-3	Green	548	–	–	–	13.62	1.82	0.70	0.023	0.04	7.35	19	592	282	31

[a]λ = color tone, quantitative parameter of color in the XYZ color system.

Table 4.3 Content of Characteristic Elements-Impurities in Amazonites and Paragenetic Potassium Feldspars from Various Genetic Types of Rocks and Formations of Granitoids, g/t

Formational and genetic types of rocks	Country, region	Pb		Rb		Cs	
		Microcline	Amazonite	Microcline	Amazonite	Microcline	Amazonite
Alaskite, pegmatites: Exocontact Rare-metal-muscovite	India	280 ----------(5) 150–585	410 ---------(20) 100–900	855 --------(9) 570–1346	2100 --------(9) 900–3300	66 --------(9) 10–130	156 ---------(9) 18–280
Inner-granitic	Baikal region	32 ----------(8) 26–40	160 -------(6) 40–500	1419 ---------(8) 710–1975	1990 --------(8) 1300–2477	98 --------(7) 13–217	191 --------(7) 100–312
Subalkaline-leucogranite: Amazonitic granites	Transbaikal	32 ----------(6) 14–60	160 ----------(5) 60–250	589 ----------(10) 427–954	3100 --------(31) 1410–5000	12 --------(7) 8–15	32 --------(8) 23–50
Vein derivatives of granites	Kazakhstan	54 --------(6) 18–100	450 ---------(6) 150–810	1870 --------(4) 1150–2880	5800 ---------(14) 3800–7900	17 ---------(18) 10–22	64 ---------(8) 24–120
Ongonites: phenocrysts	Mongolia	113 --------(10) 28–360	98 ---------(22) 53–250	3094 ---------(9) 1815–4415	5007 ---------(25) 3958–6400	40 -------(7) 17–82	109 ---------(20) 40–170
Matrix	Mongolia	62 ----------(2) 51–74	77 --------(18) 12–320	706 ----------(4) 84–1500	2290 ---------(18) 1000–3000	–	–
Sodium-lithium pegmatites	Kola Peninsula	–	–	11000 --------(43) 5500–24000	–	1000 --------(4) 600–2400	–
Alkaline-granite REE pegmatites	Kola Peninsula	120 ----------(34) 50–250	3410 --------(103) 300–9300	800 --------(21) 500–1000	3400 ---------(43) 1100-6800	28 -------(16) 10–50	320 --------(38) 30–600
Rare-metal pegmatites	Urals	90 --------(12) 40–180	460 --------(47) 50–1100	960 ---------(10) 560–1500	7730 ---------(62) 1000–22000	60 ---------(10) 20–90	270 ---------(15) 30–530
Endocontact metasomatites	Ukraine	30 ----------(7) 10–50	210 --------(7) 80–400	2380 ---------(10) 930–3830	2900 ---------(7) 1000–4200	22 ---------(10) 16–34	34 ---------(3) 24–48

N.B.

1. The table is composed according to materials of the author et al. [47], as well as of V.S. Antipin, V.A. Dunaev, K.K. Zhirov et al.
2. In the numerator are indicated mean values, in the denominator the extreme, and in the parentheses, the number of analyses.

Table 4.4 Levels of Concentration of Typical Elements-Impurities in Amazonites and Microclines from Pegmatites of Various Type, g/t

| Type of pegmatite, region | Mineral | Evaluation of mean concentration | | | | | K/Rb | K/Cs | K/Pb | Rb/Cs | Rb/Pb |
		Pb	Rb	Cs	U	Th					
REE, Kola Peninsula	Amazonite	3360 ± 820	2880 ± 320	258 ± 45	12.5 ± 3	27.2 ± 4	39.8	760	76	18	1.9
	Microcline	150 ± 20	695 ± 40	21.7 ± 2	<2	<2	157	7487	1878	54	38.8
Rare-metal, Urals	Amazonite	375 ± 30	5820 ± 670	264 ± 30	11.4 ± 2	13.9 ± 3	32.7	1191	316	42	10.3
	Microcline	85 ± 6	940 ± 120	52 ± 6	<2	<2	100.5	2250	1117	23	15

N.B. In the table are calculated the maximally probable evaluation of mean concentration (and error of its determination at 5% of the levels of magnitude), which in the opinion of D.A. Rodionov and V.V. Ivanov, is more effective with the use of the data of different representativity (literature and particular analytical materials).

Table 4.5	Content of Secondary Elements-Impurities in Amazonites and Microclines from Pegmatite Deposits, g/t			
Element	Kola Peninsula		Urals	
	Microcline	Amazonite	Microcline	Amazonite
Tl	20 ---------(14) 4–70	34 ---------(54) 10–150	8 ---------(5) 0–10	17 6–30 ---------(15)
Ga	35 ---------(12) 5–66	60 -------(58) 30–150	72 ---------(4) 10–100	120 --------(10) 40–250
Be	3 --------(14) 1–5	10 ---------(24) 2–20	5 ---------(8) 3–7	9 --------(8) 5–20
Y	–	20 --------(50) 10–150	–	–
Ba	350 --------(10) 200–520	135 ---------(18) 10–200	200 ---------(5) 100–300	100 ----------(6) 10–270
Sr	180 ----------(5) 100–300	56 --------(24) 30–200	125 -----------(6) 100–200	11 ---------(10) 2–28

N.B. In the numerator are indicated mean values, in the denominator the extremes, in the parentheses the number of analyses.

color, expressed in the replacement of blue potassium feldspars by green ones (of various tonality). The latter, in terms of setting within the vein bodies as well as in terms of correlations with amazonites of other colors, belongs to the very latest generations (Photo 14), in which have been reported the greatest concentrations of potassium and minimal sodium.

An analogous tendency is observed in amazonitic pegmatites of the Il'menskie Mountains: in samples of amazonites from the central parts of the bodies—where pegmatite of pegmatoid and blocky textures is developed—a minimal content of albitic ingrowths is noted. Moreover, potassium feldspars and amazonites from the peripheral (fine-to medium-grained) and intermediary (graphic) zones always contain more albite in the ingrowths, and consequently are richer in sodium, than do amazonites constituting the large pegmatoid and blocky formations in the central parts of the vein (Table 4.2). It follows to note that in the Il'menskie pegmatites have been observed as it were an inverse to the aforementioned correlation of variously colored amazonites: particular to the later generations are predominantly blue and greenish-blue colors. However, it needs to be kept in mind that the green colors of the Il'menskie amazonites represent an optical mixture of blue and yellow/red-brown colors of potassium feldspars.

Deperthitization is highly distinctly manifested in amazonitic granites as well. Potassium feldspars from the pluton of the subalkaline-leucogranite formation during the process of their amazonitization are purged of perthitic ingrowths of albite: the most intensively colored amazonites in the vein genetic types and metasomatites of granites embedded in the endo- and exocontacts of the plutons are practically devoid of albitic ingrowths. It is interesting to note that such wide variations of contents of sodium and potassium were noted already in 1970 by V.V. Gordienko likewise in potassium feldspars of rare-metal (sodium–lithium) pegmatites.

2. The known situation is confirmed that the potassium feldspars associated with amazonites but themselves of common (non-amazonitic) colors always are depleted of lead, rubidium, and cesium. On the contrary, elevated concentrations of these elements are an inherent feature of the chemical composition of amazonites, although in various genetic types of rocks of granitoid formations accumulates a varying quantity of the noted element-impurities (Table 4.3). Thus, in the Kola amazonites have been observed contents of lead (excluding amazonite-orthoclase from the Broken Hill deposit) that are the maximum possible in potassium feldspars; therefore, the mean value of contents of this element here is at its highest. In amazonites from the principally rare-metal (Il'menskie) pegmatites are noted the highest contents of rubidium.

It follows, however, to emphasize that according to the data of N.A. Solodov, the contents of rubidium as well as of cesium that are the maximum possible for potassium feldspars have been reported not in amazonitic varieties, but in microclines with common colors from several types of pegmatites of the sodium–lithium type. Unsurprising, therefore, are the high contents of rare alkalis in the weakly colored amazonites characteristic for the separate pegmatite types of the subalkaline-leucogranite formation.

A low isomorphic capacity for impurity elements distinguishes amazonites from pegmatites of the alaskite and subalkaline-leucogranite formational types. A position average in terms of the content of characteristic element-impurities is occupied by amazonites from the ongonites and pegmatoid formations located in the endo- and exocontacts of granites of the subalkaline-leucogranite formation.

3. Chemical analyses of amazonites indicate that their composition contains the constant presence only of one chromophore element, iron, the content of which varies from $0.0n$ to $0.n$ wt% (in conversion to Fe_2O_3). Furthermore, certain differences have been established in the contents and correlation of di- and trivalent iron in amazonites with different

colors. There is no doubt that both valent forms of iron play a defining role in the composition of amazonitic feldspars, in which the content of di- and trivalent iron is commonly close to that found in certain microclines from the ceramic and amazonite-containing pegmatites.

4. First indication of the high contents in amazonites of radioactive impurity elements—uranium and thorium (Table 4.4). Those data were obtained by M.N. Ostrooumov in 1976 and have been verified only for pegmatites of the alkaline-granite formation [13].

As has already been noted, the maximum intensity of amazonitic coloration is characteristic for pegmatites of the alkaline–granite formation, in which, as a rule, have been reported the greatest concentrations of rare and trace elements in intensively colored amazonites. We consider in greater detail certain particularities of the chemical composition of amazonites from a range of pegmatite deposits for this formational type.

Rubidium and Cesium. Rubidium was uncovered in the Il'menskie amazonite in the quantity of 3.12 wt% as early as 1913 by V.I. Vernadsky, and multiple times in the most recent works, other researchers have confirmed the presence of elevated concentrations of this element and of cesium in amazonites of various deposits.

The analysis of the obtained samples indicates that in amazonites content of rubidium as a rule is of an order higher than in microclines from pegmatites of the mica-bearing-ceramic type. At the same time, contents of rubidium and cesium similar to those of amazonite are particular to potassium feldspars with common colors from rare-metal pegmatites (of sodium–lithium type), which, according to the data of N.A. Solodov, are distinguished from amazonites only by their contents of lithium and cesium that are the maximum possible for potassium feldspars.

However, in amazonites from various deposits, rubidium is accumulated to an irregular degree. Thus, the average concentration of rubidium in amazonites from pegmatites of Precambrian age (Kola Peninsula) is approximately half that calculated for amazonites from the youngest late Paleozoic pegmatites (see Tables 4.3 and 4.4).

Therefore, in the latter, the accumulation factor for rubidium (7–8) is higher than in Kola amazonites (3–4). An inverse ratio of the accumulation[1] factor in amazonites of these deposits is noted for cesium (approximately 11–12 in Kola and 5–6 in Il'menskie pegmatites). The same can be said for concentration[2] factors of these elements in amazonites. Statistical verification of the hypotheses on the equality of average values of contents of rare alkalis permits (with probability of 95%) comes to the conclusion that in terms of the concentrations of these in amazonitic potassium feldspars from Kola and Il'menskie deposits differ substantially

---

[1]The relation of average content of the element in amazonites and this value in paragenetic potassium feldspars with common colors.
[2]The relation of average content of the element in amazonites of its Clark abundance ratio in pegmatites, according to N.A. Solodov.

from one another. Rejected also is the hypothesis of an equality of dispersions of contents of rubidium and cesium (with a low variation factor of these characteristics). If the criteria developed by V.V. Gordienko for the manifestation of a type of pegmatites and metallogenic specialization of a pegmatite province are applied to the obtained values of contents of rare alkalis in amazonite, then the following explanation results: Il'menskie pegmatites correspond in terms of content of rubidium and cesium to the spodumene type; Kola pegmatites in terms of content of cesium are also close to the latter (in terms of concentration of rubidium, they fall into the field of muscovite-feldspathic rare-metal pegmatites, according to the terminology of V.V. Gordienko). According to the X-Y plot K/Cs–Na$_2$O, proposed by those same researchers for prognosis evaluation of pegmatites, Precambrian and Phanerozoic amazonitic pegmatites are defined as rare-metallic. The values of the distribution coefficients of rubidium and cesium between potassium feldspars and micas in amazonitic and rare-metal pegmatites are likewise close, but in the latter, potassium feldspars are relatively rich in yet another alkaline element: lithium. In amazonites, content of lithium is found to be at the same level as that in microclines of mica-bearing-ceramic pegmatites; that is, approximately 1 g/t (gram per ton), in rare cases 10 g/t. In view of the limited isomorphic capacity of potassium feldspars for lithium, its discovery in amazonites as well as in common microclines can be explained by the presence in those minerals of submicroscopic particles of lithium-containing minerals (e.g., mica).

The geochemical kinship between potassium and rubidium (and to a lesser degree with cesium) enables the dispersion of rare alkalis in potassic minerals; moreover, in the opinion of N.A. Solodov, this regularity is manifest with particular distinction during the pegmatitic process. The noted tendency is traced also in amazonitic pegmatites: rare alkaline elements are accumulated in potassium feldspars and amazonites in parallel with the increase in them of content of the orthoclasic component. Furthermore, the values of the indicatory ratio K/Rb and K/Cs on the whole decrease with the transition from peripheral toward the central zones of pegmatite bodies, that is, the contents of rubidium and cesium in amazonites located in the inner parts of the vein bodies increase not only in terms of absolute value, but also in terms of their ratio to the content of potassium (see Tables 4.1 and 4.2). Highly indicative is the enrichment in rare alkalis of the very latest formations of varieties of green color, in comparison with the earlier amazonites to which are particular blue tones of color. In the change of the ratio of Rb/Cs in terms of thickness of pegmatite bodies, no regularity is observed that would provide evidence for an approximately uniform dispersion of these trace elements in potassium feldspars of amazonitic pegmatites.

Lead. The elevated content of lead in amazonites was first noted in the work of I. Oftedal in 1957. Precise quantitative measurements subsequently carried out by K.K. Zhirov and other researchers have confirmed that amazonites contain concentrations of this element.

For a comparison of the character of the distribution of lead in feldspars from various mountain rocks, we will analyze the known data from the literature.

According to V.V. Ivanov et al. (1973), see [13] the average content of lead in potassium feldspars of granitoids and pegmatites consists respectively of 46 and 52 g/t with variation from 1 to 660 g/t. According to the results of the author et al. [47], the average concentration of this element in amazonites exceeds by 4–50 times the mean values reported by the mentioned researchers. Significantly less (by 10–20 times in comparison with amazonites) lead is found in microclines from rare-metal and mica-ceramic pegmatites (V.V. Ivanov et al., N.A. Solodov). Particularly rich in lead are amazonites from the pegmatites of the Kola Peninsula, in which the highest concentrations of this element have been established: 10,000 g/t. The maximum value of the accumulation factor of lead (29–30) is also found in these amazonites (see Tables 4.1–4.4). In certain pegmatite bodies, the quantity of lead in amazonites is found to be within the range 0.03–1.0 wt% with a high dispersion of mean values and a variation coefficient close to 100%.

On a lesser order of value is the average content of lead in amazonites of the Il'menskie pegmatites; the accumulation coefficient in these accordingly decreases to 2–2.5. Similar values of contents of lead in amazonites of pegmatite and hydrothermal origin were obtained in 1965 by K.K. Zhirov and S.M. Stishov: according to the data of these researchers, in amazonites are concentrated from 50 to 2800 g/t of lead, whereas in microclines this concentration ranges from 70 to 280 g/t; in colored (amazonitic) sections of the same samples, lead content is always 2–3 times greater than in non-colored sections. The sample average and dispersions of contents of this element in amazonites of the Kola and Ural deposits differ significantly from one another.

Within the limits of separate pegmatite bodies, contents of lead differ in amazonites from different zones, with greater concentrations always found in samples from pegmatoid and blocky textures. Furthermore, the ratios of K/Pb and Rb/Pb distinctly decrease toward the center of the veins, where lead is evidently found in greater contents than are rare alkaline elements (see Table 4.1). We turn our attention to the nonuniform concentration of this element in variously colored amazonites. In blue and greenish-blue varieties, its concentration (1000 g/t) is always much lower (sometimes to an order of magnitude) than in amazonites of saturated green colors. It is indicative that green amazonites distinguished by maximum lead contents are confined as a rule to the central parts of pegmatite bodies.

Considering the above, it can be concluded that lead is a typical impurity element of amazonitic feldspar, where it accumulates and reaches the very highest concentrations unknown in other varieties of feldspars.

Uranium and Thorium. The evaluation of these elements in potassium feldspars and amazonites has been carried out by the X-ray fluorescence method in the laboratory of the Russian Geological Institute (VSEROSGEI). In total, more than 80 samples were analyzed. The processing of the results of the analyses indicates that in the studied amazonites, the average content of uranium consists of 12.8 g/t (with variations from 4 to 23 g/t), and of thorium of 22.8 g/t

(5.9–67 g/t)—significantly exceeding the average (according to V.V. Ivanov et al.) level of concentration of these elements in potassium feldspars from granitoids and various types of pegmatites. Compared with microclines from mica and ceramic pegmatites, in amazonites the content of uranium is on the same order of magnitude, while content of thorium is 16 times greater. Within the limits of separate pegmatite fields (Kola Peninsula, Il'menskie Mountains), the concentration of uranium and thorium occurs in the central parts of the vein bodies; moreover, non-amazonitized potassium feldspars from these pegmatites contain by comparison with amazonites 4–7 times less uranium and 5–8 times less thorium (concentration of these elements in microclines is found to be on the level of 1–2 g/t). Drawing attention is the varying level of concentration of thorium in amazonites of variously aged pegmatites: in Precambrian pegmatites, its average content consists of 27.9 g/t; whereas in late Paleozoic pegmatites, it is approximately two times less. However, the hypothesis on the equality of these averages is not disproved: at the same time, the average contents of uranium in amazonites from such pegmatites have almost identical value (respectively 12.5 and 11.4 g/t).

It would appear that the established fact of anomalous concentration of uranium and thorium in amazonites is not accidental, although it is probably still early to speak of an isomorphic form of their existence in the crystalline lattice of potassium feldspars. The consistently elevated contents of these elements in amazonites (in associated potassium feldspars they are always less), along with the occurrence of the highest concentrations of uranium and thorium precisely in those amazonites in which are also noted maximum concentrations of typical "amazonitic" elements (lead, rubidium, and cesium), provide evidence of a concentration of radioactive element-impurities in potassium feldspars with amazonitic colors (see Table 4.4).

Other Elements. With the aid of quantitative optical spectrographic analysis, it has been established that, compared to paragenetic potassium feldspars, amazonites are rich in thallium, gallium, beryllium, and yttrium and poor in barium and strontium (Table 4.5). The dimensions of ionic radii and the valence of thallium and rubidium are equal, in consequence of which, according to N.A. Solodov, their behavior during the pegmatite process is fully congruent. The average contents of thallium in amazonites of both Kola and Ural pegmatites are 1.5–3 times greater than in microclines with common colors from the same deposits. While it is known that amazonite is a mineral-concentrator of thallium, the greatest quantity of thallium, as well as of rubidium, has been established, according to the data of V.V. Ivanov et al., in common microclines of rare-metal pegmatites.

Content of gallium in amazonites somewhat exceeds its average concentration in potassium feldspars of rare-metal pegmatites; the values obtained by the author et al. [47] exceed by 4–8 times the average content of this element in potassium feldspars of the pegmatites enumerated by V.V. Liakhovich (14.9 g/t).

Beryllium and yttrium are typical impurity elements of amazonites from Kola pegmatites. Their accumulation occurs in amazonites inclined toward the inner zones of pegmatite bodies in which are commonly located rare earth element (REE) minerals associated with amazonite.

A two- to threefold enrichment of beryllium and a total absence of yttrium are characteristic for amazonites from pegmatites of the Il'menskie Mountains.

As is known, potassium feldspars of late generations of pegmatites are distinguished by their decreased contents of barium and strontium. Amazonites, as the very latest formation in pegmatite bodies, contain a minimal quantity of these elements—commonly 2–11 times less than in the common potassium feldspars associated with them. In the studied amazonites has been noted an approximately fourfold decrease in the concentration of barium by comparison with the average value for potassium feldspar from the pegmatites calculated by V.V. Liakhovich.

Greater contradictions are seen in the behavior of strontium: in amazonites from alkaline-granite pegmatites, its content is 3–11 times lower than in paragenetic microclines and other potassium feldspars of pegmatites. Furthermore, the (to a significant degree) radiogenic nature of strontium in amazonites should be kept in mind. Thus, according to the data of V.A. Dunaev (1967), see [1] in amazonites from the central parts of pegmatite bodies of the Il'menskie Mountains, the content of radiogenic isotope ^{87}Sr comprises from 34% to 78% of the general quantity of strontium.

It is highly indicative that intensively colored Kola amazonites contain on average five times more strontium than do Il'menskie amazonites. It is interesting to note that in terms of age, Precambrian and Phanerozoic amazonite-containing pegmatites are distinguished from each other by approximately the same factor. Additionally, within the boundaries of one of the pegmatite bodies (Kola Peninsula) has been observed an increase in content of strontium in amazonites by measure of the strengthening of its color (from the peripheral zones toward the central parts of the vein).

The seemingly contradictory behavior of this element (on the one hand, relatively decreased content in amazonites by comparison with associated non-amazonitic potassium feldspars; on the other, the undoubted elevation of its concentration in connection with the duration and intensity of the amazonitization process) can be explained by the presence of two forms of this element (non-radiogenic and radiogenic) and by the connection of the intensity and tone of amazonitic color with the concentration of the radiogenic isotope of strontium. In conclusion, we emphasize that microclines and amazonites from the Kola and Ural pegmatites in terms of average values of the noted (other) elements (with the exception of barium and strontium) do not have significant differences.

In overall summary, it can be stated that the greatest fundamental geochemical particularity of amazonite is the consistent enrichment of this mineral by comparison with associated non-amazonitized potassium feldspars in lead, rubidium, uranium, and thorium. At the same time, the quantitative evaluation of the degree of concentration, limits, and average concentration of rare and trace elements in amazonites from various deposits provides evidence of a regular change of chemical composition of this mineral, connected with the geological conditions of their formation (belonging to certain granitoid formations, geological epochs, finds, and places of localization in various genetic types of rocks, etc.). For example, from the peripheral zones to the central part of the pegmatite veins of all granitoid formations, amazonite changes its composition systematically in terms of some major (e.g., K_2O, Na_2O) and trace components (e.g., Pb, Rb, U, Th) that can be seen to increase in the amazonite samples.

4.3 STRUCTURAL STATE

Having rather firmly taken root in the petrographic-mineralogical literature is the understanding of amazonite as the maximum microcline: the most highly ordered and triclinic variety of potassium-sodium feldspars. Furthermore, certain researchers consider the maximum degree of structural order of potassium feldspars to be the necessary condition for development of amazonitic color. However, in connection with the discovery and research of new deposits and genetic types of amazonite-containing rocks, results have appeared that provide evidence of a significant variation of structural types of potassium feldspars with amazonitic color. For example, among amazonites in plutons of subalkaline (fluorine–lithium) granites and their subeffusive analogues (ongonites) as a rule have been established typical maximum microclines as well as less-ordered varieties: intermediary triclinic orthoclases and pseudo-cross-hatched twinning microclines.

But, separate finds of amazonites uncommon in terms of structure have been known even earlier: as early as 1959, in pegmatites from the Eastern Alps, A. Alker discovered amazonites characterized by inconsistent optical constants and above all a significantly varying degree of the angle of optical axes. Subsequently, H. Makart and A. Preisinger detected in these pegmatites amazonites with extremely low (including null) values of the magnitude of X-ray triclinity (Δr)[3]. Later, in pegmatites of the Broken Hill ore deposit in Australia was found a green potassium feldspar, which after detailed X-ray and optical investigations was identified as a typical orthoclase: $-2V = 61°$, $\Delta r = 0.0$; Δo (optic triclinity) $= 0.0$.

Here, it is appropriate to recall that already in the previous century A. Des Cloizeaux had divided all crystals of amazonite into two groups: (1) the "triclinic" system, with highly opaque color and low transparency; and (2) the "monoclinic" system, partially colored and more transparent (see Chapter 1). These observations did not find explanation in their own time and, apparently, were subsequently forgotten. The findings of A. Des Cloizeaux in the present time have been confirmed by the results obtained via X-ray and optical methods—alongside triclinic coarse-cross-hatched twinning intensively colored amazonites are known varieties with low intensity of amazonitic color that are distinguished by pseudo-cross-hatched twinning and a low degree of structural order.

---

[3]$\Delta r_{131,\,1\bar{3}1}$; the degree of Al/Si order according to Goldsmith and Laves (1954), see [14].

The structural varieties of amazonitic and common potassium feldspars (in all were studied approximately 200 samples) have been defined principally on the basis of the measurement of X-ray (X-ray triclinity Δr, the degree of monoclinic order Δt)[4] and to a lesser degree of optical (optical order 2V and optical triclinity Δo) characteristics, while accounting for perthitic structure and degree and type of twinning. For a range of typical amazonitic potassium feldspars has been established an order by method of infrared spectroscopy; parameters were calculated for unit cell and content of aluminum in tetrahedral positions. Diffractometric investigations were conducted in the X-ray laboratories of the St. Petersburg Mining Institute, the universities of Nantes, Mainz, and Mexico (M.N. Ostrooumov), and Geological Institute VSEROSGEI (T.A. Sosedko).

Considering the variety of amazonite's geological setting and the necessity of a differentiated approach to the examination of the geological conditions of the finds of this mineral, we will pause first and foremost on the general characteristics of the structural particularities of the studied samples [14]. Analysis of all data known from the literature and the experimental study of the author provides evidence of the following (Tables 4.6 and 4.7).

1. In pegmatites of the alaskite and alkaline-granite formations that formed in the Precambrian, amazonitic potassium feldspars belong to the maximum microcline with distinctly cross-hatched twinning and micro- and macroperthitic structure. Gross composition of amazonite-perthites fluctuates within the interval Or_{60-91}, Ab_{9-40}, while the separate phases are represented by almost purely final members (Or_{95-99}, Ab_{98-99}). These facts are situated in full accordance with widespread views, while the known settings are only somewhat detailed. The only and still not fully understood exception is the amazonite-orthoclases from the Broken Hill deposit. On the other hand, in the younger and less deep pegmatites of Paleozoic and Mesozoic age from those same formations, the dimensions characterizing the degree of Si-Al order of amazonites vary within a wider range by comparison with the practically stable values of the structural state of amazonites from ancient pegmatite rocks. The latter circumstance has been confirmed by the data in the literature (H. Makart, A. Preisinger) in terms of the structural state of amazonites from pegmatites of Alpine age. Further confirmation was obtained after the authors had available to them samples of amazonites from the Paleogenic pegmatites of the central Pamirs; among them were found pseudoperthitic amazonites, the structural particularities of which permit their classification as monoclinic orthoclases ($\Delta r = 0.0$; $\Delta t = 0.58$).

2. The greatest number of structural types has been established for amazonitic potassium feldspars from subeffusive rocks, granites, and their other genetic types of the subalkaline-leucogranite formation. Characteristically,

---

[4] $\Delta t = T_1 - T_2$; the degree of monoclinic order according to Stewart and Wright (1974), Krivokoneva and Karaeva (1989), and Altaner and Kamentsev (1995), see [46].

Table 4.6   Structural-Optical Characteristics of Amazonites from Various Genetic Types and Formations of Granitoids

Formational and genetic types of rocks	Country, region	Composition of amazonite	Δr	Δt	-2V, angle degree	Character of perthites and twinning	Structural type
Alaskite, pegmatites: Exocontact rare-metal muscovites	India	Or 70–91 Ab 9–30	0.92 ---------(6) 0.88–0.96	0.97 --------(6) 0.94–1.0	68–82	Macro-microperthites lattice	Maximum high microcline
Endo- contact granitic	Baikal region	Or 60–80 Ab 30–40	0.96 --------(12) 0.92–1.0	0.97 --------(12) 0.94–1.0	78–84	Macroperthites lattice	Maximum high microcline
Subalkaline leucogranite: Amazonitic granites	Transbaikal	Or 57–95 Ab 5–43	0.73 ---------(18) 0.2–0.93	0.92 --------(18) 0.85–0.96	68–78	Macro-microperthites, Micro-pseudolattice	Intermediary triclinic orthoclase, intermediary and maximum microcline
Other genetic types of granites	Transbaikal	Or 88–95 Ab 5–12	0.95 --------(20) 0.91–1.0	0.97 --------(20) 0.93–1.0	78–84	Micro-pseudoperthites Lattice	Maximum microcline
Ongonites:	Mongolia	Or 70–91 Ab 9–30	0.79 --------(12) 0.1–0.86	0.78 --------(16) 0.6–0.9	68–84	Micro-pseudoperthites pseudolattice	Intermediary triclinic orthoclase, Intermediary and maximum high microcline
Alkaline-granite: REE pegmatites	Kola Peninsula	Or 62–85 Ab 15–38	0.95 ---------(32) 0.9–1.0	0.97 --------(32) 0.92–1.0	80–86	Macroperthites lattice	Maximum high microcline
Rare-metal pegmatites	Urals	Or 63–90 Ab 10–37	0.91 ---------(20) 0.78–0.96	0.95 --------(20) 0.84–1.0	76–85	Macro-microperthites Macro-microlattice	Maximum high microcline
Endocontact metasomatites	Ukraine	Or 91–98 Ab 2–9	0.97 --------(5) 0.94–1.0	0.98 --------(5) 0.96–1.0	78–80	Microperthites pseudolattice	Maximum high microcline

N.B.
1. The table is comprised according to materials of the author et al. [47], as well as of V.S. Antipin, O.A. Bespal'nyi, and others.
2. In the numerator are indicated average values, in the denominator the extremes, in the parentheses the number of analyses.

these amazonite-containing rocks were formed predominantly in the Mesozoic (and to a lesser degree in the Paleozoic) and at comparatively low depths. Precisely within the granitic plutons almost all researchers have noted the appearance of a strengthening of intensity of color in tandem with an increase in degree of structural order of amazonites, a

Table 4.7 Structural State and Color of Amazonites and Microclines from Certain Pegmatite Deposits

Location, vein	Textural zone of the pegmatite, number of the sample	Color	Structural state		Distribution in tetrahedral positions			λ (nm)
			Δr	Δt	$T_1 0$	$T_{t,}m$	$T_2 0 - T_2 m$	
Kola Peninsula, Ploskaia Mountain, 19	Pegmatoid, 241	Blue	1.0	0.98	0.99	0.01	0.0	497
	Pegmatoid, 240	Greenish-blue	0.98	0.98	0.98	0.02	0.0	501
	Fine-blocky, 292	Grass-green	0.94	0.96	0.96	0.02	0.01	512
Kola Peninsula Ploskaia Mountain, vein 20	Blocky, 293	Dark-green	0.93	0.95	0.95	0.03	0.01	542
	Pegmatoid, 355	Pink	0.95	0.96	0.97	0.01	0.01	596
	Fine-blocky, 351	Bluish-pink	0.98	0.96	0.97	0.01	0.01	500
	Fine-blocky, 345	Blue	0.98	0.98	0.98	0.01	0.0	497
Il'menskie Mountains, mine 70	Medium-grained, 70-84-1	Yellowish	0.88	0.98	0.93	0.05	0.01	593
	Graphic, 70-84-2	Pale-green	0.82	0.98	0.90	0.08	0.01	572
	Pegmatoid, 70-84-3	Yellowish-green	0.94	0.95	0.94	0.01	0.02	569
Il'menskie Mountains, mine 77	Medium-grained, 77a	Grayish-pink	0.81	0.92	0.88	0.08	0.02	587
	Graphic, 77-2	Greenish	0.82	0.86	0.88	0.06	0.03	570
	Pegmatoid, 77-3	Green	0.96	0.98	0.97	0.01	0.01	548
Pamirs, Murgab region, Mika vein	Pegmatoid, 501	Pale-green	0.0	0.56	0.38	0.38	0.12	553
	Pegmatoid, 502	Bluish-green	0.0	0.71	0.42	0.42	0.08	503

finding that serves as the basis for the currently accepted views on the close connection between these two processes. Needless to say, it cannot be denied that the structural-optical states of potassium feldspars with different intensities of specific amazonitic color change in correlation to their localization in different zones of plutons; furthermore, within the latest-established endo- and exocontact pegmatoid and hydrothermal veins are formed as a rule maximally ordered and intensively colored amazonites.

3. In ancient Precambrian pegmatites, notable differences in terms of structural state have not been established between amazonites and their paragenetic potassium feldspars of common colors; at the same time, as they are in the relatively young and less deep rocks (pegmatites, granites, and ongonites), amazonitic varieties on the whole are always more ordered. Common potassium feldspars in pegmatites of Precambrian and Paleozoic age belong to the maximum high microclines, whereas in Meso- and Cenozoic pegmatites, granites, and subeffusive rocks, they are represented

principally by non-ordered varieties: monoclinic and triclinic orthoclases and high and intermediary microclines ($-2V = 57$–$76°$; $\Delta r = 0.0$–0.8; $\Delta t = 0.56$–0.92).

We pause now on the characteristics of the structural state of amazonites and potassium feldspars from concrete pegmatite deposits and granitic plutons belonging to the various formational types.

In the Precambrian pegmatites of the alkaline-granite formation (Kola Peninsula), the structural state of common potassium feldspars as much as that of weakly and intensely colored amazonites is characterized by a high degree of order and in terms of all measured X-ray–optical parameters corresponds to typical maximum high microcline (Δr, $\Delta t > 0.9$; $2V > 80$). In terms of the values of Δr, Δt, $2V$, Δo, and the infrared order between amazonitic and non-amazonitic feldspars from the same vein bodies, notable differences are not observed. Most importantly, the structural-optical constants of common potassium feldspars fall within the interval of the corresponding characteristics of the structural state particular to amazonites. This, nonetheless, does not imply that all amazonites in terms of structural state are fully analogous to common potassium feldspars; the latter can be even relatively more ordered.

Given a normal course of evolution of perthites and twinning of potassium feldspars of pegmatites, the tendency of their structural order predominates. As indicated by the research of the author et al. [47] conducted in the Ploskaia Mountain district, in amazonitic pegmatites, the noted tendency can be violated. Thus, in the very latest formation, amazonites that coincide principally with the central zones of pegmatite bodies and are characterized by green colors, a heterogeneous twinned structure is observed under the microscope. Furthermore, there are delineated sections, which in terms of dimensions of twinned sub-individuals and measure of optical triclinity differ substantially from one another. Simultaneously, early amazonites of blue color with large planar ingrowths of albite are composed of twinned individuals fairly close in terms of dimensions. This phenomenon provides evidence of minor structural disorder and is connected with the fact that low-temperature amazonites, by comparison with high-temperature varieties, as a rule contain the maximum concentrations of such cations large in terms of dimensions as rubidium, cesium, and lead. According to V.V. Gordienko, the inclusion of rare alkaline elements in the position of the potassium feldspathic lattice enables the stabilization of a disordered monoclinic modification, the content of which increases with the growth of contents of rubidium and cesium. Characteristically, in such potassium feldspars alongside the triclinic is present also a monoclinic phase, revealed by the X-ray method to contain rubidium and cesium components together of no less than 1%.

According to the data obtained through interpretation of the X-ray diffraction pattern, not one of the samples researched by the author (not even the varieties with the greatest contents of rubidium and cesium—up to 2.5% rubidium and cesium components) represent a combination of triclinic and monoclinic modifications; that is, they all consist only of the triclinic phase. Additionally, late amazonites without a doubt possess a definite degree of structural disorder (the values of Δr and Δt in late generations are always lower: see Table 4.7). Thus, it can be proposed that the concentration of rare alkaline elements (and possibly lead) affects only the decrease in degree of structural order of amazonites, while it is insufficient for the formation in their structure of the monoclinic phase.

Considering that late generations of amazonitic potassium feldspars are colored predominantly in green hues (in distinction from early blue and greenish-blue varieties), we are led to the logical implication that a deepening of color of amazonites (replacement of blue potassium feldspars by green ones of various tonality) corresponds to a decrease in degree of structural order. There is no question of any connection between the latter and the intensity of color, insofar as even monocrystals with immediate or gradual transitions of common feldspar into amazonite are reported to contain a practically identical (within the margin of error of the measurement of the value of Δr and Δt) degree of order of various parts of mineral individuals. Finally, the existence of a direct connection is called into doubt by an already noted fact: certain varieties of amazonites are distinguished by a lesser degree of structural order by comparison with paragenetic potassium feldspars of non-amazonitic colors.

In late Paleozoic pegmatites of the alkaline-granite formation (Il'menskie Mountains) by measure of proximity to the central zones of the vein bodies is increased the quantity of amazonite and the intensity of its color. In the same direction, there is a general increase in the degree of structural order of potassium feldspars as well. In terms of structural parameters, various mineral individuals of the latter are fairly distinct: microclines with common colors ($\Delta r = 0.79-0.89$; $\Delta t = 0.86-0.92$; $-2V = 74-75°$) differ markedly from pale-colored amazonitic varieties ($\Delta r = 0.89-0.94$; $\Delta t = 0.95-1.0$) and particularly from amazonites ($\Delta r = 0.82-0.98$; $\Delta t = 0.93-1.0$; $-2V > 82°$).

According to B.M. Shmakin (1968), see [14] within the limits of the separate crystals, amazonitic sections are characterized by markedly great values of the magnitude of X-ray triclinity by comparison with uncolored sections, comprising 0.7–0.8 for the latter and 0.95–1.0 for the former. Regarding the multicolored monocrystals measured by the author that are localized in the central parts of the vein bodies in which have been observed vague borders and gradual transitions between the variously colored zones, marked differences have not been established in the structural state of parts of individuals heterogeneous in terms of color (in yellow and white parts of crystals $\Delta r = 0.9-0.94$; $\Delta t = 0.93-1.0$; in amazonitic zones the values of these parameters are located within the same range). The same can be said for monocrystals with sharp borders between the uncolored and amazonitic-colored sections. Finally, we refer to those multicolored individuals in which microcline twinning has been observed macroscopically in yellow as much as in green sections of monocrystal.

As has already been noted, amazonite in pegmatites is confined to the various textural zones (graphic, apographic, and pegmatoid) of the vein bodies, distinguished by intensity and tone of bluish-green color. Moreover, among the very latest formation intensively colored amazonites have been uncovered the greatest ordered as much as relatively disordered varieties (see Table 4.7). It is indicative that in the latter have commonly been established the maximum concentrations of rare alkaline elements and lead. Thus, following the study of samples from Ural pegmatites, the understanding has been formed of the non-singular character of the connection between amazonitic color and the degree of structural order of feldspar.

Analogous regularity is revealed in the large, not infrequently zonal plutons of amazonite-containing granites of the subalkaline-leucogranite formation. The general particularity of such granites is the consistently reported increase in degree of amazonitization in direction toward the pre-dome zones (see Section 2.3.1). The most intensively colored have been found to be the potassium feldspars in the hydrothermal veins and metasomatites of granites localized in endo- and exocontacts of plutons.

Biotitic granites embedded in the "inner zones" and in the deep horizons of such plutons are composed of non-amazonitic potassium feldspars distinguished principally by a low degree of structural order ($-2V = 65–70°$; $\Delta r = 0.0–0.6$), permitting their classification as intermediary order orthoclases and microclines. Occasionally, however, in these parts of separate intrusions are found potassium feldspars with high values of Δr. As N.V. Gerasimovskii and N.V. Zalashkova have suggested, the presence of relatively ordered lattice of microcline ($\Delta r = 0.81–1.0$) is characteristic for ore-bearing biotitic granites, whereas intrinsic in the majority of cases to potassium feldspars of non-ore-bearing plutons is a low magnitude of X-ray triclinity—$\Delta r = 0.0–0.2$. Unfortunately, in limiting themselves solely to the certification of facts, these researchers did not consider the reasons for a phenomenon of such importance from the prospecting point of view.

In amazonitic granites are commonly identified several generations of potassium feldspars, including amazonites. According to the data of Ia.S. Kosals, amazonites of early generations are distinguished from those of later generations by the character of the deposits, an indistinct cross-hatched twinning, lesser idiomorphism, and a low intensity of color. Equally commonly occurring among the first generations of this mineral, which are localized primarily in the "intermediary zones" of the pluton, are ordered ($-2V = 80–83°$; $\Delta r = 0.85–0.91$) as well as disordered varieties (with lower values of those parameters). In endocontact sections of a range of plutons of Kazakhstan and Transbaikal have been uncovered amazonitic potassium feldspars with a maximum ordering of the structure and pale colors.

According to the data of P.V. Koval', in the endocontact zone of one of the plutons of Transbaikal composed of quartz-amazonitic rock of pegmatoid texture, transparent cross-hatched twinning amazonites of blue color, belonging to the early generations, are characterized by the magnitude of Δr of 0.2–0.3. These amazonites are, thus, more disordered by comparison with the still earlier potassium feldspars with reddish-brownish colors, for which the magnitude of Δr has changed within the interval of 0.1–0.75. In the monocrystal of these potassium feldspar, by measure of transition from blue section to the pink section, has been reported a growth in the magnitude of Δr from 0.35 to 0.71.

Amazonites located in vein derivatives of granites are completely devoid of such variations of values characterizing the structural state. In these amazonites have been observed only highly ordered varieties ($-2V = 85–88°$, $\Delta r = 0.9–1.0$; $\Delta t = 0.95–1.0$) with color changing in terms of intensity and tonality. According to the data of the author, in the zonal potassium feldspars from quartz-amazonitic veins situated in endo- and exocontacts of plutons, sections colored in bluish-green and white color share an identical degree of structural order. The reported examples quite demonstratively confirm the understanding of the non-singular character of the connection between structural state and colors of the examined potassium feldspars in amazonite-containing granite plutons. The data received during the process of the research have led to review once more the direct connection that has been postulated (L.A. Ratiev, Kh.N. Puliev, B.M. Shmakin et al.) between the degree of manifestation of amazonitic color and the structural order of potassium feldspars. Here, it follows to turn attention to the fact that of all the research that consider the variants of the influence of ordering on the color of amazonite, not one cites any proof of a direct influence of migration of aluminum to a definite structural position on the coloration of potassium feldspar with bluish-green hues of color.

Numerous experiments conducted in 1956 by E. Spencer, Iu. Goldschmidt, and F. Laves on the study of the thermal stability of various potassium-sodium feldspars have furnished evidence that the lattice structure of microcline-perthites disappears only after extended (500–700 h) heating at temperature no lower than 1000 °C. In the experiments of T. Bart, certain microclines, even after exposure to these temperatures over the course of 300–600 h, remain triclinic. Finally, there are the commonly known results of Iu. Goldschmidt and F. Laves, who in examining the correlation of the stability of microcline and sanidine, researched in detail the activity of dry annealing on microcline and discovered its high thermal stability (Table 4.8). The annealing of this mineral over the course of several dozen up to 100–150 h at temperatures of 800–1000 °C does not cause a sharp disorder of its structure. On the contrary, extended annealing at 1000–1065 °C causes a gradual disordering of structure of microclines up to the transition to the monoclinic modification. Moreover, an uninterrupted decrease is noted in the value of the angle of the optical axis and of X-ray triclinity.

With the aim of verifying the resulting contradictions, M.N. Ostrooumov has conducted specialized experiments for studying the change under heating and X-ray irradiation in the structural state of potassium feldspars (including amazonites) from the principal genetic types of amazonite-containing rocks examined above.

We note the general particularities of thermal decoloration of amazonitic color. During a series of experiments performed with constant time and variable annealing temperatures, the loss of color begins at 200–300 °C, and its full and rapid disappearance occurs within the interval of 450–500 °C. The curves of thermal decoloration, according to M.N. Ostrooumov, provide evidence that the

Table 4.8	Change of the Factor of Triclinity Δr with High-Temperature Annealing of Microclines from Pegmatites			
Conditions of annealing		**Δr**		
Temperature (°C)	Time (h)	Initial	After annealing	Location
1050	48	0.96	0.92	Switzerland
1050	144	0.96	0.77	Switzerland
1050	240	0.96	0.53	Switzerland
1050	336	0.96	0.29	Switzerland
1050	720	0.96	Monocline	Switzerland
800	144	0.84	0.84	Madagascar
1060	192	0.84	0.62	Madagascar
1060	336	0.84	0.62	Madagascar
1060	672	0.84	0.20	Madagascar
800	192	0.96	0.92	Finland
1065	140	0.96	0.74	Finland
1065	169	0.96	0.69	Finland
1065	284	0.96	0.61	Finland
1065	666	0.96	Almost monocline	Finland

first sharp drop in absorption is observed when the heating reaches 280–300 °C. Extended heating at these temperatures (over the course of 1–50 h) does not lead to any change in color, that is, such a temperature regime did not manage to fully "remove" amazonitic color; the final raising of the temperature to 400–450 °C in the curve of thermal discoloration corresponds to a second fall in absorption. All studied samples of amazonites at temperature of approximately 500 °C were discolored over the course of 0.1–0.2 h.

In result, through the study of structural characteristics at all stages of heating (from the lowest temperatures at which decoloration begins up to the temperatures at which the mineral is discolored practically instantaneously), the above-indicated temperatures and exposure times were demonstrated to be clearly insufficient for fundamental Si-Al disordering of amazonites (and more so for the transition to the monoclinic modification).

The results cited in Table 4.9 provide confirmation that the parameters of the structural state of decolored varieties do not substantially differ from those of the initial samples. Even high-temperature annealing conducted over the course of relatively short time does not lead to fundamental disordering of the structure. The cross-hatched twinning of amazonites, likewise, remains unchanged after heating. These observations provide a sufficient basis for concluding that with decoloration of amazonites no change occurs in the distribution of aluminum in terms of various tetrahedral positions in the structure. In precisely the same way, after X-ray and gamma-irradiation of the preliminarily decolored samples, no fundamental change is noted in the degree of ordering either in varieties that have picked up artificial color or in samples that were not colored under the influence of hard X-ray irradiation (Table 4.10).

Table 4.9 Parameters of the Structural State of Amazonites after Heating at Various Temperatures and over Times of Exposure

Number of sample, color of amazonite	Conditions of annealing		Initial value		Value after annealing		Location, type of rocks
	Temperature (°C)	Time (h)	Δr	Δt	Δr	Δt	
241, Blue	200	5	1.0	0.98	0.98	0.95	Kola Peninsula, pegmatite
241, Bluish	300	5	–	–	1.0	1.0	Kola Peninsula, pegmatite
241, Pale-blue	400	5	–	–	0.97	0.94	Kola Peninsula, pegmatite
241, White	500	26	–	–	0.94	0.90	Kola Peninsula, pegmatite
295, Green	200	5	–	–	0.96	1.0	Kola Peninsula, pegmatite
295, Greenish	300	5	–	–	0.96	0.98	Kola Peninsula, pegmatite
295, Pale-green	400	5	–	–	0.95	0.96	Kola Peninsula, pegmatite
295, Grayish-white	500	26	–	–	0.94	0.91	Kola Peninsula, pegmatite
111, Dark-green, after annealing white	1000	1	0.98	0.88	0.78	0.82	Il'menskie Mountains, pegmatite
70-1, Dark-green, after annealing yellowish-white	500	26	0.96	0.96	0.93	0.85	Il'menskie Mountains, pegmatite
70-86, Bluish-green, after annealing white	500	26	0.96	0.91	0.93	0.84	Il'menskie Mountains, pegmatite
268, Blue, after annealing white	500	26	0.97	0.94	0.94	0.92	Transbaikal, granite

N.B. Samples 241 and 295 are represented by a series of cleavage plates selected respectively from blue and green monocrystals. They were characterized by an identical initial structural state and subsequently annealed at different temperatures and times of exposure (cited here are only a part of the obtained results; color is indicated for the heat-treated samples).

Table 4.10 Results of X-ray Structural Analysis of Amazonites

Number of sample, characteristics	Structural state		Distribution in tetrahedral positions			Parameters of lattice cells						
	Δp	Δz	T_10	T_2m	T_20-T_2m	a (Å)	b (Å)	c (Å)	V	α	β	γ
241, Initial blue	1.00	0.98	0.99	0.01	0.00	8.597	12.955	7.218	721.73	90.5	116.0	87.8
241, Decolorized, white	0.96	0.96	0.97	0.01	0.01	8.595	12.974	7.218	723.16	90.8	115.9	87.5
241, Irradiated, pale-blue	0.96	0.99	0.96	0.02	0.01	8.593	12.971	7.218	723.09	90.7	116.0	87.6
295, Initial green	0.96	1.00	0.97	0.01	0.01	8.594	12.971	7.221	723.23	90.9	115.9	87.6
295, Decolorized, white	0.94	0.98	0.96	0.02	0.01	8.600	12.971	7.219	723.83	90.7	116.0	87.6
295, Irradiated, grayish-white	0.96	0.96	0.99	0.01	0.00	8.602	12.966	7.218	722.59	90.5	116.0	87.8

N.B.

1. Values of Θ for all *samples* equal to 1.00.
2. Thermal processing of amazonites was conducted in the laboratory oven at a temperature of 500°C over the course of 5h; decolorized samples were irradiated with a [60] Co dose of 1.3 MeV.
3. Analyses were performed in the X-ray laboratory of the University of Nantes by M.N. Ostrooumov; the conditions of the study: Siemens 5000 diffractometer, Co-radiation, 30 kW, 15 mA, recording rate of 0.02 deg/min, external standard of NaCl, parameters of lattice cell were calculated by the least-squares method on a Pentium 4 computer using the program JCPDF2004 (PDF4).

According to the experimental data of the author et al. [47], the exposure of amazonites, even over the course of 30–50h at 1000 °C, did not lead to the type of changes in the value of Δr and Δt that are characteristic for monoclinic potassium feldspars (the values of these magnitudes in amazonites subjected to such a heating regime are respectively 0.4–0.6 and 0.6–0.9). Analogous results have been obtained by H. Makart and A. Preisinger [14]. According to their data, the magnitude of Δr of amazonites from various deposits shifts from 0.95 to 0.99 (with the exception of a sample from the pegmatites of the Eastern Alps, where $\Delta r = 0.0$–0.1). After heating to 1000 °C over the course of 48h, the value of Δr of the researched samples decreased insignificantly from 0.92 to 0.94 (the value of Δt of the amazonites did not change). Significant disordering at such exposure temperature is reported in amazonites that have undergone extended annealing (no less than 80–100h), which is entirely insufficient for obtaining fully disordered monoclinic potassium feldspar (after such thermal treatment $\Delta r = 0.2$–0.3; $\Delta t = 0.5$–0.6).

The obtained results are based not only on the measurement of values of X-ray triclinity, degree of monoclinic and infrared-ordering, and angle of optical axes, but also on the data of full X-ray structural analysis (see Table 4.10). It, thus, becomes clear that despite widespread understanding, the maximum degree of ordering of potassium feldspars is not a necessary condition for the development of amazonitic coloration. The above-cited geological observations and corresponding experimental results provide evidence that amazonitic color can be possessed by potassium feldspars with an entire range of structural states: from low orthoclase to high-ordered microcline. On the other hand, it cannot be denied that the most intensively colored (typical) amazonitic potassium feldspars from various genetic and formational types of granitoid rocks are characterized by a predominantly high structural ordering, though among minimally ordered varieties of amazonite such intensive color is unknown.

This question will be examined in greater detail in Section 5.2. Here, we note the following: the absence of disordering with the disappearance of color, along with finds of amazonites with a lowered degree of structural order and of monoclinic potassium feldspars with amazonitic color indicate that the primary impurity elements that form defect centers of color are not connected with the aluminosilica framework, but they are situated within the spaces of the lattice in the place of alkaline cations. The low intensity of the color of disordered amazonites, which are represented as a rule by early high-temperature generations, is explained by the insignificant concentration of impurity elements and their concomitant centers of color, and is in no way connected with the migration of atoms of aluminum to a defined position of the crystal lattice of potassium feldspar.

Late and relatively low-temperature generations of amazonites that have been subjected to decomposition, ordering, and twinning, are characterized by the most intensively expressed specific color due to the presence in them of the highest concentrations of element-impurities that participate in the formation of color-forming centers. The absence of a direct link between the appearance of the color of amazonite and the phenomenon of structural ordering of potassium feld-spars does not exclude the possibility of the existence of an indirect correlation between those and other factors (the influence of Si-Al order on the distribution of alkaline cations and impurity elements at various structural positions).

4.4   COLOR AND ITS SPECTRAL-COLORIMETRIC RESEARCH

4.4.1   Colorimetry

Color represents the defining sign (or property) of amazonite that permits its identification as an intraspecies variety of potassium-sodium feldspars. In the textbook and guidebook literature, the name "amazonite" is commonly applied to a green or bluish-green variety of microcline without providing a distinct understanding of the actual color of this stone. The great diversity of its color has long been noted in the geological-mineralogical literature. Thus, M.P. Mel'nikov (one of the first researchers of the Il'menskie mines) already in 1882 wrote, "Amazonstone rarely is of apple-green color or varies up to turquoise color, but commonly is yellowish-grey with green flaws; upon closer examination, it is visible how its surface is mottled by yellowish-grey streaks, through which its color changes." In a similar way, using a touch of artistic description, amazonite color was later characterized by A.E. Fersman.

As has already been noted in Chapter 1, the first instrumental evaluation of the color of amazonites in terms of the parameters of the curves of spectral absorption was performed in 1949 by E.N. Eliseev. Subsequently, spectromet-ric measurements of amazonitic color were gradually broadened [8,12,19] and acquired a necessary and basic character in works treating amazonite [9,20]. At the same time, alongside spectrometry, the method of colorimetry began to be applied in the evaluation of the variability of colors of amazonite [10,15,46].

The nuances of the color of amazonite, previously attracting the attention of only a narrow circle of mineralogists, is at the present becoming an object of specialized research. Reasons for this new interest are found in the following circumstances. First, the grade and value of an amazonite as an ornamental stone are dependent on the color, and in part on its tone and saturation. Second, the fine particularities of color depend on the crystal chemical features of amazonite and correspondingly on the conditions, mode, and time of its for-mation and transformation [10]. Third, the necessity noted first by the author for a differentiated approach to amazonite as an important prospecting criteria [47], confirmed subsequently by the renowned specialist in the field of applied mineralogy A.I. Ginsburg and by other specialists on rare-metal mineralization, requires the study of the specifics and varieties of its color.

Thus, important in the scientific as well as the practical regard is the quantita-tive objective evaluation of the color of this stone, based not only on its spectral characteristics but also on the spectral particularities of the human eye and the source of light, whereby occurs the definition of color of amazonite [15]. These primary factors of influence on the evaluation of the color of mineral objects are most fully accounted for by the international XYZ colorimetric system. The precision of the calculation of the color of minerals in this system, on the one hand, is based on the use of experimentally measured coefficients of spectral reflection or transmission in the visible range, and on the other hand, on account of the color-sensing particularities of the visual perception (uniformly tabulated values) with any of the standard sources of light. This system has been described in fair detail in the specialized literature already accepted in mineralogical practice and, therefore, will not be characterized in detail here; we will only note that in result of uncomplicated analyses are obtained three colorimetric parameters: the quantitative characteristic of the color—the lightness— Y; and the qualitative—chromaticity factor—x, y. The latter are plotted on the standard color chart (the color triangle, Fig. 4.6), on which, with the aid of specially composed nomograms, is carried out the transition to a characteristic color (chromaticity) of the researched mineral objects—obtaining values of the color tone λ (hue) and the saturation (chroma or purity) of color P. Thus are measured and classified the quantitative characteristics of the color of minerals in the form of colorimetric parameters x, y, Y, λ, and P, calculated with one of the standard sources of light, A, B, C, and D_{65}, used in the present system.

At the contemporary stage in colorimetry are widely accepted three instrumental methods of measuring color: spectrophotometric measurement, calibration,

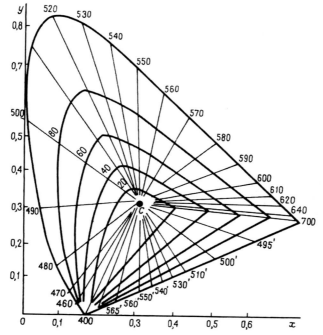

FIGURE 4.6
Color chart of the XYZ colorimetric system.

and direct measurement. In the majority of cases, preference is given to the first of the indicated methods, insofar as it is distinguished by the maximum degree of objectivity of the conducted measurements, which, moreover, are performed with the aid of commercial equipment listed in the official register of color-measuring instruments. Finally, by comparison with the other methods, spectrophotometric measurement commands yet another merit very important for mineralogists—its calculation of color is conducted on the curve of spectral reflection (transmission), which is a highly informative characteristic in the research of the nature of mineral color.

Focusing attention on the above, it is necessary to note that the color characteristics of amazonite cited below have been obtained by the spectrophotometric method (spectrophotometers and spectrometers UV–VIS–NIR of various models for the verification of comparability of results) in the standard conditions of measurement in a series of uniformly prepared samples—on a plate polished along [001] with an area of 20×20 mm. Spectral coefficients of reflection of the researched samples were read from the corresponding curves across 10 or 20 nm and were entered in a computer for calculation of all colorimetrical parameters via a specially composed Basic program. Selected in the capacity of a standard source of light was C_{1961}, the spectral composition of which corresponds to the diurnal phases of blue light and approximately simulates conditions of common natural lighting.

In the present section are laid out the results of a colorimetric study of amazonites from the primary geological formations of this mineral: pegmatites of the alkaline-granite formation, as well as granites and their other genetic types of the subalkaline-leucogranite formation. The examination primarily of colorimetric, rather than spectral, characteristics of amazonitic color is connected with the fact that in the following chapters they are involved in the explanation of the various changes in optical spectra of amazonites. In all were researched approximately 300 samples from the principal deposits of Russia situated predominantly in the Kola Peninsula, the Urals, and Transbaikal, as well as from Kazakhstan (Tables 4.11 and 4.12).

Table 4.11 Colorimetric Parameters of Amazonites from Various Genetic Types and Formations of Granitoids

Formational and genetic types of rocks	Region	Age	Colorimetric parameters		
			λ (nm)	P (%)	Y (%)
Subalkaline-leuco-granite: Amazonitic granites	Kazakhstan	Hercynian	522–575	5–10	44–65
Subalkaline-leuco-granite: Amazonitic granites	Transbaikal	Cimmerian	497–570	5–10	48–70
Alkaline–granite pegmatites: REE	Kola Peninsula	Proterozoic	491–545	10–25	25–42
Rare-metal	Urals	Hercynian	495–570	5–20	49–69

Number of sample	Color of amazonite (visual characteristics)	Region, deposit, rock	Colorimetric parameters		
			λ (nm)	P (%)	Y (%)
25	Blue	Kola Peninsula, western Keivy, pegmatite of blocky texture	491	22	33.1
40	Greenish-blue		495	17	28.5
19	Bluish-green		500	17	29.4
35	Grass-green		512	20	30.7
5	Emerald-green		522	17	27.1
24	Dark-green		530	18	31.5
39	Tobacco-green		543	21	28.9
330	Turquoise	Southern Urals, Il'menskie Mountains, pegmatite of pegmatoid texture	495	15	18
70-1	Yellowish-green		564	20	45
70-59	«–»		558	16	50
327-1	«–»		574	14	61
70-10	Green		548	18	50
77	«–»		542	15	52
59	Greenish		510	6	69
70-84-2	Pale-green	The same, graphic pegmatite	575	17	53
77-2	«–»		574	22	53
20-3	Greenish-blue	Transbaikal, medium-grained granite	500	5	54
1-3	«–»		500	7	70
5-3	Greenish		550	5	65
6-3	Green		560	5	57
3-3	«–»		520	5	63
28-M	«–»	Kazakhstan, medium-grained granite	512	5	47
23-M	«–»		550	5	48
31-M	Greenish		570	9	61

Table 4.12 Colorimetric Parameters of Amazonites from Concrete Pegmatite Deposits and Granite Plutons

In correspondence with the obtained data was established the existence of a fairly wide area occupied on the standard color diagram in the triangle of chromaticity by parameters of colors of amazonites (Fig. 4.7). The colorimetric characteristics of this mineral turned out to be on the whole fairly mutable, varying within the following range: color tone λ=491–575 nm; purity (saturation) of color P=5–25%; lightness Y=25–70%. Furthermore, however, there can be noted a definite distinction of colors of amazonites from the various deposits (see Table 4.12). Thus, the greatest purity of color characterizes amazonites from the pegmatite deposits of the Kola Peninsula, and the least purity characterizes the amazonites from the granite plutons of Kazakhstan and Transbaikal (Fig. 4.8).

Opposite correlations are noted for those same samples in terms of values of lightness. A certain intermediary position in terms of these parameters is occupied by amazonites from the pegmatites of the Il'menskie Mountains. The diagrams constructed within the coordinates λ–P and λ–Y show a marked

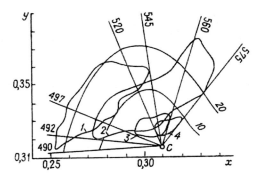

FIGURE 4.7
Coordinates of coloring of amazonites. (1–4) Contours of the field of coloring of amazonites from pegmatites of the alkaline-granite formation (1: Kola Peninsula, 2: Il'menskie Mountains, Urals) and granites of the subalkaline-leucogranite formation (3: Transbaikal, 4: Kazakhstan).

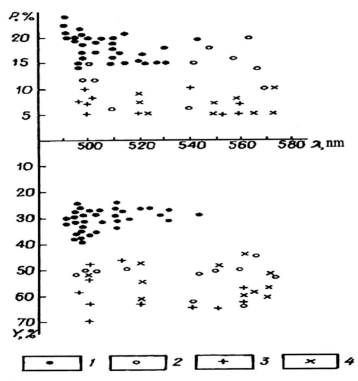

FIGURE 4.8
Colorimetric parameters of amazonites in coordinates of the purity P, color tone, λ and lightness Y. Amazonites from pegmatites (1: Kola Peninsula, 2: Il'menskie Mountains, Urals) and granites (3: Transbaikal, 4: Kazakhstan).

separation of the fields of colorimetric parameters of the Kola amazonites, at the same time as the analogous characteristics of these minerals from other deposits to a significant degree overlap one another.

In Chapter 2, it was noted that amazonite with colors that are the maximum possible in terms of intensity is highly typical for the most ancient (Proterozoic) pegmatites of the alkaline-granite formation, whereas in the fairly young (Jurassic–Cretaceous) plutons of the subalkaline-leucogranite formation, amazonite of pale colors is particular to granites and the metasomatites connected with them. In terms of the calculated colorimetric parameters of amazonitic colors, on account of the data on the variously aged series of granitoid rocks, a noted tendency in the general form is not difficult to formulate: to the younger and less deep formations of various genetic and formational types are inherent amazonites with lesser purity and greater lightness of color and a somewhat greater interval of values of magnitude of color tone. In other words, in the young series of granitoid rocks has been noted an idiosyncratic weakening (dilution) of amazonitic color. When granitoids of single-type formations of variously aged series are examined, the same tendency turns out to be expressed most of all in the rocks of the alkaline-granite formation. This, in part, is supported by recent finds of pale-colored amazonites in the alkaline-granite pegmatites of Alpine age.

An increase in purity and a lowering of lightness of color with an insignificant widening of the range of values of color tone occurs in amazonites by measure of the transition from relatively early to the later (younger) formations inside the series. This is particularly distinctly established for ancient series of granitoid rocks.

An analogous regularity is noted likewise with the transition from early generations of amazonite to late and from rocks of early phases to late.

The most diverse zonality and sectoriality of distribution of color are observed within the range of monocrystals of amazonitic potassium feldspar. Thus, in the latter, certain pyramids of intumescence commonly are colored in hues varying in terms of colorimetric parameters; others are more frequently colorless or white. Not uncommon are cases where from the center to the periphery of crystals, amazonitic color becomes progressively saturated and dark (occasionally even with a change of color tone) and, conversely, fades to the point of the manifestation of colorless (white) outer zones.

A similar multifaceted character of the change of chromaticity of amazonites from early to late members of evolutionary ranges supports the necessity of a differentiated approach to this mineral as a prospecting indicator. Judging by the preliminary data, amazonites with fully defined colorimetric parameters are found in various deposits in paragenetic associations with ore mineralization. This suggests the possibility for employing the colorimetric characteristics of amazonitic feldspar in the evaluation of the prospects of certain types of deposits connected with rare-metal-bearing granitoid formations.

Thus, when judging according to objective colorimetric parameters, it is not difficult to arrive at the result that the most saturated (pure) colors are possessed

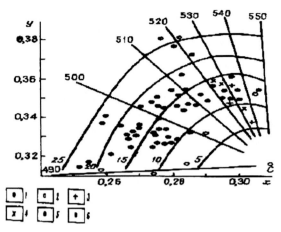

FIGURE 4.9
Coordinates of coloring of amazonites from the Ploskaia Mountain pegmatite deposit (1) and other typical green and blue gemstones (2: turquoise, 3: nephrite, 4: jadeite, 5: malachite, 6: emerald).

by amazonitic potassium feldspars from alkaline-granite pegmatites of the Kola Peninsula. It is evidently not incidental that pegmatite deposits precisely of this region serve at the present as the principal source of amazonite. In connection with this, it follows to focus attention on the colorimetric characteristics of amazonite of the Ploskaia Mountain deposit, which now is the primary source of this mineral. In amazonites of the given deposit, the values of color tone are distinguished by the greatest variations (Fig. 4.9); the change of these values within the 491–545 nm interval corresponds to the transition from blue to green tones of color through an entire series of intermediary varieties. In this same direction decrease the purity and lightness of amazonitic color (color somewhat degrades). The obtained results fairly simply explain the irregular consumer relationship toward amazonite of the Ploskaia Mountain deposit—it is known that on the market blue and transitioning to green varieties of this mineral have seen and continue to see much higher demand. Finally, there is yet another distinguishing particularity of this deposit: it is practically devoid of the yellowish-green and greenish-yellow varieties of amazonites so typical, for example, for the Il'menskie pegmatites.

Geological surveying and exploitation works on the Ploskaia Mountain deposit conducted at different times and by different specialists visually have differentiated an irregular and usually small number of color varieties of amazonite. The author et al. [47] with a selection of samples for colorimetry have described no less than 50 visually differing (in terms of color) amazonites. It is indicative that characteristic for each of these varieties are an individual pattern, form, and content of perthitic ingrowths. Established likewise is a defined spatial and temporal localization of color of amazonite in the given deposit: in the central and lower parts of the vein body are concentrated predominantly green varieties of amazonite. Connected in close spatial proximity to certain of these are rare-earth and rare-metal accessory mineralizations. With this concrete example, it is not difficult to ascertain, on the one hand, the significant arbitrariness of traditional verbal-visual evaluation of amazonite color without the corroboration of objective measurements, and on the other, the actuality of specialized colorimetric research of this stone.

Relying on the obtained colorimetric parameters of amazonites of the given deposit, there is no difficulty in dividing them in terms of values of color tone into four groups: $491-494\,nm = blue$; $495-500\,nm = greenish-blue$; $501-510\,nm = bluish-green$; and $511-545\,nm = green$. Within these color groups, differences in colors of amazonites are defined by degree of saturation (purity) and lightness. The greatest number of color differences is established in the last group. It is interesting that many of them in terms of colorimetric parameters (Photo 21, see Fig. 25) are highly close to such semiprecious stones as emerald, malachite, turquoise, jadeite, and nephrite [9,23]; if the first two stones, as a rule, exceed amazonite in terms of purity of color, then nephrite and jadeite in terms of this parameter in the majority of cases even are somewhat inferior to

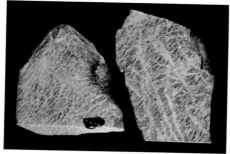

PHOTO 21
Amazonites of turquoise ($\lambda = 492\,nm$, $P = 20\%$, $Y = 34\%$) and emerald-green ($\lambda = 525\,nm$, $P = 19\%$, $Y = 32\%$) colors.

PHOTO 22
Blue amazonite with zonal distribution of perthitic ingrowths of albite with a single tone of color ($\lambda = 494\,nm$) and different purity and lightness. (Ploskaia Mountain deposit, western Keivy, Kola Peninsula).

amazonite. Precisely in this way, certain blue varieties of amazonites in terms of purity of color are comparable with better samples of turquoise. Thus, support is found for the relevance of comparing colors of amazonites with the color of the most known and utilized blue and green semiprecious stones.

In conclusion, we indicate those circumstances that from the point of view of the author are most important for the evaluation of color of amazonites. The color of this mineral (as of any other non-transparent and optically irregular semiprecious stone) in practice is evaluated in reflection light, in which connection the most objective of its characteristics will be the spectral curve of reflection in the visible range of the spectrum. With that, however, it must be kept in mind that the obtained spectrum of reflection is not always sufficiently selective, given the conditionality of not only the particularly amazonitic potassium feldspar, but also of other phases (in part albitic ingrowths), which are an inherent component of practically any amazonitic feldspar. Perthitic ingrowths render a twofold influence on the colorimetric parameters of amazonitic color: the increase of their quantity leads to a decrease in purity and a growth of lightness of color of amazonite.

It follows to note, as well, the evaluation of color of multicolored amazonites. In this case may be obtained an integral as much as a local characteristic of color. With local change of spectra of reflection of monocrystal sections varying in terms of color, commonly reported are greater values of purity and lesser values of lightness with variation of the value of color tone from 560 to 570 nm. Integral evaluation of color leads to the opposite phenomenon: a decrease in purity and a growth of lightness, as well as to the long-wave shift of values of color tone (hue) in the range of tones transitional from green to yellow (575–580 nm).

The selectivity of optical spectra of this mineral can be increased by means of surveying the spectra of transmission of separate translucent micrograins. Such spectra are necessary for researching the nature of centers of amazonite color; however, the use of these data in the calculation of colorimetric parameters appears to be from the methodological point of view incorrect or at least possible (feasible) for an isolated case of the conditions of the observation: macroscopically perceptible color of the entire sample will not correspond to such of its isolated microsection, the more so measured in transmitted light.

At the contemporary stage of development of mineralogy, a subjective verbal-visual evaluation of color, not relying on colorimetry, should be considered an anachronism. Therefore understood are the attempts by many researchers to express by "number and scale" any of the primary properties of any mineral. At the same time, in the majority of cases, color measurements of minerals are performed on noncommercial instruments, with different geometry of lighting and observation, different radiation sensors, and nonuniform standards of white surface. It is not difficult to understand that the results obtained in this way cannot be considered comparable. The further implementation in mineralogical practice of instrumental methods of measurement of color demand from researchers strict compliance with two conditions: the use of commercial color-measurement instruments and a unified methodology of experimental measurement [15].

4.4.2 Optical Spectroscopy

Amazonitic crystals and aggregates as a rule are non-transparent or in the best case are translucent through in fine plates, in consequence of which the spectrometrical measurements of amazonites performed to date have been based only on the spectra of diffuse reflection, obtained principally from the polished surface of the sample. The first measurements permitted the establishment of the participation in the formation of amazonitic color of at least two absorption bands: the first is in the short-wave domain with a maximum of ~380 nm in the nearest ultraviolet (UV) region. The second is a long-wavelength band in the visible region between 625 and 740 nm. The band in the visible range is more characteristic of amazonite [8]. Because of this, the current hypothesis regarding shades of amazonite color (from clear blue to intense greenish-blue and different shades of green) depends on the ratio between the relative intensities of the ultraviolet and visible (amazonitic) spectrum bands.

In the initial stage, works [12] in their evaluation of color with the aid of spectrum, besides the position of both coloring bands, accounted for their intensity and area or correlation of areas. According to these characteristics was noted a certain separation of amazonites from various types of rocks and formations. Thus, it was determined that amazonites from Precambrian alkaline-granite pegmatites (Kola Peninsula) are distinguished by the widest variation of spectral composition of colors (the maximum of the amazonitic band was observed within the 630–740 nm interval) and the highest values of intensity of the "amazonitic" band and the corresponding area of long-wave and short-wave absorption bands; that is, particular to these amazonites are the most distinctly manifested, particularly amazonitic colors. A less intensive and narrower spectral composition of amazonitic colors characterizes samples from the near-coeval pegmatites of Karelia. A similarly narrow spectral characteristics but within a more short-wave range (625–650 nm), as well as a highly significant variation of intensities and corresponding by the areas of the absorption bands, are inherent to samples of amazonites from late Paleozoic pegmatites (Urals, Russia; Pikes Peak, USA). The position of the maximum of the "amazonitic" band of amazonites from the pegmatoid formations in granites of the subalkaline-leucogranite formation (Transbaikal) are exceeded by the Kola and Ural samples with intensive amazonitization, close to that of the most "short-wave" Ural amazonites.

Amazonites from granites are distinguished in terms of spectral characteristics from other samples first and foremost by the lowest values of intensities of the "amazonitic" absorption band.

On the basis of the data of M.N. Ostrooumov et al. [12], conclusions have been formed on the wide variability of the characteristic parameters of spectra of amazonites, as well as on a certain specificity of color of this mineral from various

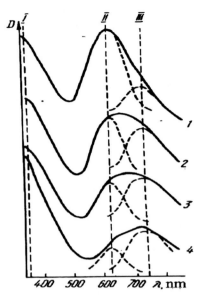

FIGURE 4.10
UV–VIS–NIR spectra of the studied amazonite samples. Measured spectra shown by solid lines, deconvoluted spectra shown by dashed lines. Symbols: (1) blue; (2 and 3) bluish-green; (4) green amazonite variety. The position of the wide absorption bands (I–III) in the optical spectra of variously colored amazonites. From blue (1) to green (4) changes the correlation of the intensities of the elementary bands (dash line); D = optical density.

genetic and formational types of granitoid rocks. Subsequently, the same author [9] through detailed analysis of spectra of variously colored amazonites from the pegmatites of Ploskaia Mountain established a non-elemental structure of the "amazonitic" absorption band, representing as a rule a superposition of two absorption bands with maximums of 625 and 720 nm (Fig. 4.10). This fact has permitted an explanation of the reason for the replacement of the maximum of the "amazonitic" band, consisting of a shift in the relative intensity of the 625 and 720 nm absorption bands that compose it. The independent behavior of these bands and their connection with various centers of luminescence have allowed the possibility to conclude that their natures differed.

So, the main absorption band in the visible range changes its position according to the observed color and specific color shade of amazonite. By applying the Fast Fourier Transform (FFT) Self Deconvolution technique [47], we were able to resolve two principal peaks between 600 and 750 nm, which compose the main characteristic absorption band in the spectra of natural amazonite. We established that the apparent shift of this main band from 625 in blue amazonite to 740 nm in green amazonite is due to intensity redistribution between those spectral constituents. In Fig. 4.10, the deconvoluted spectra (dashed lines) are shown, which correspond to the raw spectra of different amazonite generations. If we assume that this large asymmetric and intense absorption band could be a spectral area of some elemental components, we can show that two Lorentzians would be enough to correctly describe the experimental spectrum that was obtained. Analyzing in greater detail the spectra of different amazonite colors reveals that the main absorption band is characterized by a complex structure that, in reality, is a superposition of the two absorption bands at ~625 and ~740 nm (Fig. 4.10). The maximum displacement of this complex zone corresponds to a variation in the relative intensities of these two absorption components.

The color of the amazonite and its two principal different shades were determined, in the first place, by the existence of centers of colors that provoke two absorption bands in the region of 625–740 nm. We must point out that the second absorption band at ~740 nm is the main band in the visible spectrum of the green amazonite, whereas the band of approximately 625 nm always prevails in the spectrum of blue amazonite.

More detailed research of the optical spectra of absorption of amazonites has been performed by the author with the utilization of polarization equipment. For the review of the extensive collection of amazonites from various deposits, in the capacity of a reference point were selected two practically transparent samples (Ploskaia Mountain), sharply distinguished visually in terms of tonality of color (in transmitted light): bright-blue ($\lambda = 490.5$ nm) and bright-green ($\lambda = 520.0$ nm). Spectra of the diffuse reflection of these samples are characterized by extreme values of the λ_{max} maximums of the 625 and 720 nm absorption bands. From these samples were prepared plane-parallel plates along (010), that allowed the obtaining of polarized spectra of optical absorption along two axes of optical indices (Nm and Np) constituting an angle of 18° with crystallographic axes a and c respectively. With the study of samples in polarized light, it turned out that blue as much as green amazonite possesses sharply expressed pleochroism: from deep-blue and bright-green in Nm to pale-blue and pale-green in Np.

Optical spectra of absorption of the reference point samples of amazonites have been obtained in an automated single-beam polarized microspectrophotometer at room temperature within the 380–2500 nm range. The absorption coefficients were calculated for the thickness of the absorption layer of 1 mm. Absorption in the short-wave range of the spectrum (200–400 nm) was studied with the aid of monochromator DMR-4, hydrogen lamp, and standard acquisition module with FEN-39. Subsequently, these investigations were conducted as well in the laboratories of a range of foreign universities, the results of which confirmed the first experiments.

In the optical spectra of the researched amazonites have been revealed the following bands of absorption:

1. A group of ultraviolet bands of ~250, 285, and 320 nm; in spectra of green amazonite, the intensity of all of these bands is commensurate, while in spectra of blue, the 285 and 320 nm absorption bands have a clearly subordinate value with the dominating role of the 250 nm band (Fig. 4.11).
2. Polarized bands in the neighboring ultraviolet range of spectra (380 nm), the maximum intensity of which is observed at ε II Nm; its relative intensity differs for variously colored samples.

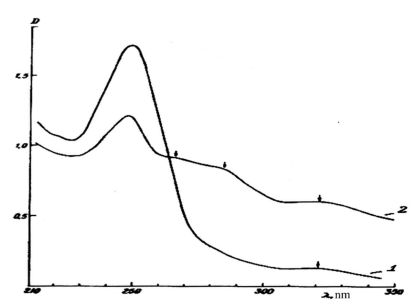

FIGURE 4.11
Absorption bands in the ultraviolet range of the spectrum of blue (1) and green (2) amazonites.

3. Polarized bands in the red range, that is, the "amazonitic" band, which in spectra of variously colored samples is characterized by specific particularities: in blue amazonite in Nm-polarization, it has a maximum of 635 nm and is practically symmetrical, but the low asymmetry of its long-wave branch (~700 nm) provides evidence of the presence of a long-wave component (auxiliary band); in Np-polarization, the maximum of the band is replaced up to value 615 nm, and the asymmetry is manifested just as weakly (Fig. 4.12); in the green sample, the "amazonitic" band is split into two components in one orientation: 650 and 725 nm (ε II Np) and is not divided in another; also noted is a wide band with a maximum 670 nm (ε II Nm); the configuration of the short-wave band supplies the basis for proposing its non-elementary character (Fig. 4.12).

4. A weak polarized band of ~1000 nm, more distinctly expressed in the spectra of blue amazonite.

5. A weak narrow band of ~420, 450, 475, 500 nm, occasionally manifested in the form of rungs on the long-wave branch of the UV-band of absorption.

Within the 1000–2500 nm spectral range, an absorption band could not be identified.

As has been noted, amazonites are fully decolored with heating up to 500–600 °C. From the series of experiments conducted with constant time and variable temperature of heating, it follows that the color of the studied samples markedly fades beginning at a temperature of 250–300 °C, and fully and rapidly disappears in the interval of 450–500 °C.

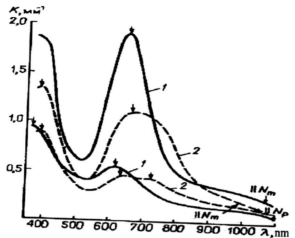

FIGURE 4.12

Polarized optical spectra of the absorption of blue (1) and green (2) amazonites in the visible and near-infrared ranges of the spectrum (K = absorption coefficient).

Variously colored amazonites are characterized with other conditions being equal by a heterogeneous thermal stability of color: with the same values of temperature a great speed of decoloration is particular to fundamentally green amazonites, the optical spectra of which are distinguished by a significant role of the absorption band in the 700–750 nm range.

Through experiments on the graduated (exposure of 10 min, interval 100 °C) thermal decoloration of amazonites was established the following:

1. The heating of blue amazonite up to 200 °C leads to a certain elevation of intensity of the 625 nm band, and at temperature of greater than 300 °C is noted a marked decrease in the intensity of the "amazonitic" band (at 500 °C by three times) and of short-wave (400 nm) absorption (Fig. 4.13(a)). If the duration of exposure of sample at 500 °C is increased to 1 h, then the "amazonitic" band is observed in the form of weak flat maximum. It follows to note that by measure of the decrease in intensity of the "amazonitic" band with the thermal processing of the sample occurs a distinct replacement of its maximum in the short-wave range— from 635 nm (in Nm-polarization) in the initial sample to 620 nm in the sample heated up to 500 °C. Furthermore, in parallel with a decrease in the intensity of the "amazonitic" band in the spectrum of the heated samples are distinctly manifested weak narrow bands of absorption of 450, 475, and 490 nm, commonly not identified in spectra of initial bright-colored amazonites.

2. More fundamental are the changes in the spectra of absorption with heating of bright-green amazonite (see Fig. 4.13(b)). In the 200–300 °C temperature interval, the common intensity of the "amazonitic" band remains practically the same as in the spectrum of the initial sample; however,

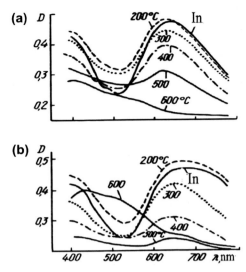

FIGURE 4.13
Change of absorption spectra of blue (a) and green (b) amazonites heated at various temperatures (200–600 °C); In = spectra of initial samples.

there is a distinct shift in its form on account of the gradual disappearance of the short-wave component in the 700–750 nm range. At 400 °C, this component is identified only in terms of the asymmetry of the short-wave branch of the "amazonitic" band, the intensity of which, as in the case of blue amazonite, decreases approximately by three times, and the position of the maximum shifts from 680 to 635 nm (in Nm-polarization). In parallel is noted a decrease in the short-wave absorption. The exposure of the sample of green amazonite at a temperature of 500 °C in the course of 1 h leads to the disappearance of the characteristic amazonitic spectrum and the appearance of secondary gray and even black color, coinciding, as is clearly visible under microscope examination, to the fracture cleavage and to other mechanical defects.

In Fig. 4.14 are cited curves of thermal decoloration of dark-bluish-green amazonite for two values of the lengths of waves: 625 and 720 nm. Insofar as the "amazonitic" band represents a superposition of two absorption bands (see Fig. 4.10), the curves of thermal decoloration by a significant measure reflect a total effect of annealing of these bands. Nevertheless, the examination of these correlations allows us to conclude that the annealing of band $\lambda_{max} = 720$ nm in the interval 20–200 °C accompanies a certain strengthening of the short-wave component from $\lambda_{max} = 625$ nm. The entire process of thermal decoloration of amazonite occurs in two steps, as is evidenced by the break in the dip of the curve in the 300–400 °C interval. The obtained data are in agreement with the author's earlier proposal mentioned above of the participation in the formation of the "amazonitic" band by a lesser measure of two absorption bands with varying

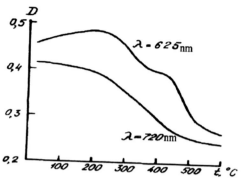

FIGURE 4.14
Curve of thermal decoloration of amazonite for various wavelengths.

thermal stability. It cannot pass unmentioned that the disruption of the centers of color, responsible for the band of $\lambda_{max} = 720–740$ nm, in the interval of temperatures of 20–200 °C, leads to the auxiliary saturation of the ~625 nm band; these circumstances provide evidence of the possible electron exchange between the various centers that cause the color of the amazonites.

It follows especially to note that with a decrease in intensity of the "amazonitic" band during the heating of samples occur fundamental changes in the UV-range of the spectrum. The 285, 320, and 380 nm absorption bands fully are annealed before 500 °C, in consequence of which the absorption of amazonites in the 260–400 nm interval sharply decreases. At the same time, a marked increase is observed in the intensity of the narrow 250 nm absorption band.

Colorimetric research of samples of variously colored amazonites (in transmitted light) heated up to various temperatures (from 100 to 600 °C) has shown that the process of their thermal decoloration is accompanied by a decrease in the values (by 2–3 times) of purity (saturation) of color; the color tone practically does not change.

M.N. Ostrooumov has conducted a range of experiments on the restoration of amazonitic color in thermally decolored samples by means of their X-ray irradiation and subsequent thermal experiments [47]. The results of radiation-thermal treatment were reported according to the change of spectra of diffuse reflection of the powdered (size of grains 0.1–0.25 mm) samples prepared from the above-described samples. For an objective evaluation of the character of the change of color, colorimetric parameters were calculated. In the spectra of the X-ray-irradiated preliminary thermally decolored initial blue ($\lambda = 491$ nm) and green ($\lambda = 502$ nm) amazonites is manifested a diffuse absorption with widely diffused maximum in the 550–620 nm range (Fig. 4.15). In the spectrum of the blue sample against its backdrop is expressed fairly distinctly an auxiliary weak absorption band with a maximum 520 nm; blue amazonite with this maximum in the spectrum after decoloration and irradiation acquires a dark-bluish-purplish-gray color ($\lambda = 447$ nm), and green amazonite acquires a purplish-gray

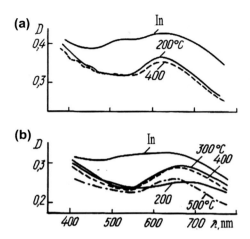

FIGURE 4.15
Change of absorption spectra of blue (a) and green (b) amazonites with secondary radiation-induced color heated at various temperatures (200–500 °C); In = spectra of samples with initial radiation-induced color.

color ($\lambda = 454$ nm). The irradiated color is distinguished from the natural also by lowered values of purity (by several times) and lightness (by several percentages).

Thus, merely direct irradiating treatment of the thermally decolored amazonites could not obtain even a weak purely amazonitic color, which is clearly identified visually as much as through the aid of objective measurement—in terms of the character of the spectrum of diffuse reflection and in terms of colorimetric parameters. The principal particularity of the irradiated samples is a purplish-gray color, and of their spectra a diffuse, practically nonstructural absorption in the entire visible range—a low maximum of reflection in the dark-blue-violet range is responsible also for the purple (violet) tonality of color.

Experimentally, it has been proven that "amazonitic" color is partially reproduced with extended thermal processing of irradiated samples. Thus, in the spectrum of the initial blue amazonite secondarily heated after irradiation to 200 °C, the wide diffuse bands (see Fig. 4.15(a)) corresponding to the purplish-gray component of the color disappear, while an absorption band in the 625 nm range appears; the sample acquires a distinctly bluish-green color ($\lambda = 504$ nm), practically not changing with extended heating up to 300 and 400 °C. The purity of this color is quite low (several percentages), while the lightness is somewhat higher (by 3–5 wt%) than in the irradiated sample. With heating up to 500 °C, the color markedly weakens, and at 600 °C, it fully disappears.

In a somewhat different fashion, does the color of the initially green irradiated amazonite change with heating? With heating to 200 °C, likewise disappearing from its spectrum is the diffuse absorption responsible for the purplish-gray color, while a weak wide band of absorption with maximum 700 nm appears; the sample is colored with a pale-green color ($\lambda = 565$ nm). Extended heating up to 300 °C leads to a slight elevation in intensity of the "amazonitic" band, the replacement of its maximum from 650 nm, and correspondingly to a certain strengthening of color ($\lambda = 531$ nm; purity and lightness somewhat increase). Just as with heating of the initial sample, at 400 °C is effected a decrease in the long-wave component of the "amazonitic" band, and with heating up to 500 °C, there is a marked decrease in the intensity of both components of "amazonitic" color; full thermal decoloration occurs at 600 °C.

It follows especially to note that the capability to restore color in decolored and subsequently irradiated amazonites strongly depends upon the temperature and time of the first heating. The experiments indicate that already at temperatures of 700 °C amazonitic color by means of the above-described radiation-thermal method can with great effort be restored. After heating (even short-term, over the course of 10–15 min) up to 800 °C and higher, amazonite color is not restored, and the sample acquires a different saturation and a hue of yellowish-brown color.

4.4.3 Luminescence

Within a multiyear period of research, vast experimental material has been accumulated on the luminescence of impurity and particular defects in crystals of feldspars, including amazonites [20]. It has been established that all studied amazonites, independently from the particularities of their geological setting and color, acquire a marked luminescence with photo- and X-ray excitation, as well as a thermally stimulated luminescence. The spectra of luminescence of practically all samples are characterized by the presence of four principal bands of emission, distributed in the 200–1000 nm range (Fig. 4.16). Likewise of principal importance, these bands are particular not only to amazonites but also to practically all types of potassium feldspars with insignificant variations in the spectral position of their maximum.

Analysis of the data from the literature [47], which includes a fully adequate consideration of possible models of the optically activated centers responsible for these bands, has indicated clear contradictions in the interpretation of certain bands. Furthermore, the experimental study of the processes that cause the luminescence of amazonite has led to the discovery of a range of results that have not obtained adequate explanation, including emission in the infrared range, the manifestation of high-temperature peaks of thermal luminescence in consequence of thermal processing at temperatures higher than 700 °C and the subsequent X-ray irradiation of crystals of amazonite, and a range of other facts that cannot be interpreted within the existing conceptual concepts. In this connection, the demand has arisen for a synthesis of the obtained data and a review of certain models in light of new results of measurements.

Principal experiments in the study of spectra of photo- and X-ray luminescence and curves of thermal luminescence have been performed with powders and monocrystals of 130 samples of amazonites from a range of pegmatite and granite deposits. Spectra of X-ray luminescence have been measured at 300 K with exposure to X-ray irradiation of tube BSV-2 (45 kW, 20 mA) through monochromators SF-4 and ISP-51 with the aid of photomultipliers FEU-39 and FEU-108.

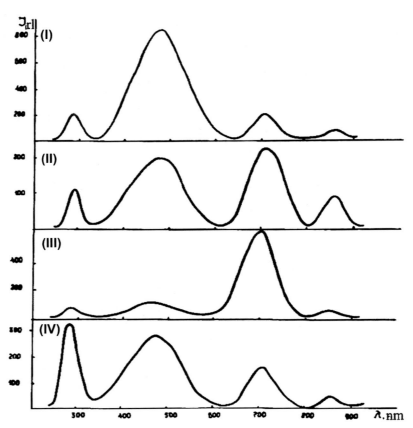

FIGURE 4.16

Typical spectra of X-ray luminescence (rl) J_H of variously colored crystals of amazonite from Hercynian (I) and Cimmerian (II) granites of the subalkaline-leucogranite formation, as well as REE (III) and rare-metal (IV) pegmatites of the alkaline-granite formation.

These measurements have permitted the establishment of the following:

1. At room temperature, the spectra of X-ray luminescence of all samples consist of four principal radiation bands: an ultraviolet band with a maximum ~285 nm (half-width ~0.4 eV); two in the visible range of the spectrum—an intensive wide band with $\lambda_{max} = 480$–490 nm (~0.7 eV) and a narrow band with $\lambda_{max} = 710$ nm (0.2 eV); and a narrow intensive band in the infrared range with $\lambda_{max} = 860$ nm (half-width 0.2 eV). See Fig. 4.16. The appearance in the spectrum of other bands (in part, 320–340, 390–420, 570, and 740 nm) is connected with the impurity of albite.

2. The intensity of the indicated bands for crystals from various deposits varies within a very wide range; however, the type of the spectrum determined in terms of its intensity remains practically unchanged for all samples within the limits of a single deposit (Table 4.13, see Fig. 4.16).

Table 4.13	Mean Values of Intensity of Bands in the Spectra of X-ray Luminescence of Samples of Amazonite from Various Genetic Types and Formations of Granitoids						
Formational and genetic type of rocks	**Region**	**Age**	**Intensity (arbitrary units) with λ max (nm)**				
			280	**480**	**710**	**860**	
Subalkaline-leucogranite: Amazonitic granites	Kazakhstan	Hercynian	31	67	300	50	
Subalkaline-leucogranite: Amazonitic granites	Transbaikal	Cimmerian	175	169	13	122	
Alkaline–granite pegmatites: REE	Kola Peninsula	Proterozoic	110	780	108	34	
Rare-metal	Urals	Hercynian	40	181	223	63	

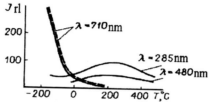

FIGURE 4.17

Dependence of intensity of X-ray luminescence of amazonite J_{rl} upon temperature in various bands of the spectrum (dashed line = with photo-excitation).

3. Temperature strongly influences the spectra of X-ray-luminescence. In the very same sample, the course of the temperature dependence is specific for each band. An increase in temperature causes a total quenching of the emission of some centers and in a range of cases stimulates the emission of other centers (Fig. 4.17). The maximum quenching action of temperature is rendered on the emission in the ~710 nm band; at 200–220 °C, the intensity of this band reaches almost null values. Drawing attention also is the fact that the temperature quenching of emission in the ~710 nm range with photo- and X-ray excitation is subordinate to a single law. It follows also to note that according to the data of A.N. Tarashchan [20], the character of temperature dependence of luminescence in the ~285 and 480 nm bands is dependent by significant measure on the impurity composition of amazonite.

4. The heating of samples of amazonite at fixed temperatures in the 400–900 °C interval variously influences the intensity of bands of X-ray luminescence (Fig. 4.18). The intensity of the 710 nm band is practically dependent on the temperature and time of heating; only at high temperatures (1000 °C) and with a duration of heating on the order of 40–50 h does the luminescence in this range of the spectrum practically disappear. The intensities of the ~285 and 480 nm bands pass through the maximum values at temperatures of 500–700 °C and eventually, by measure of the

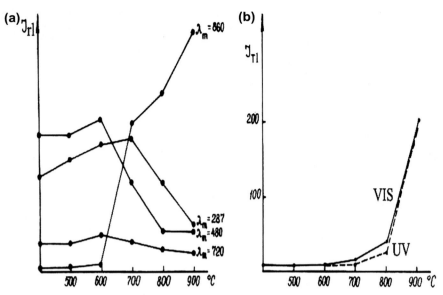

FIGURE 4.18
Influence of annealing temperature T (1 h in isothermal regime) on the intensity of bands in the spectrum of X-ray luminescence J_{rl} (a), as well as on the intensity of the peak of thermoluminescence J_{tl} with $T_{max} = 470\,°C$ in the bands with $\lambda_{max} = 285\,nm$ (UV) and $\lambda_{max} = 480\,nm$ (VIS) of green amazonite (b).

increase in temperature and time of heating, decrease insignificantly by comparison with the initial values. The most fundamental changes occur in the infrared range of the spectrum. The intensity of the ~880 nm band practically does not change in the 20–600 °C interval; however, already at 700 °C, it sharply grows (approximately by an order of magnitude greater by comparison with the natural sample) and eventually continues to grow with an increase in temperature and duration of heating. Within this temperature interval (700 °C and higher) occurs the following:

a. There is a loss of the capability of heated crystals of amazonite to restore color (with heating in the air and in the vacuum) by X- and γ-ray excitation.

b. There is a loss of the capability of heated crystals in the infrared luminescence (880 nm band) with photo-excitation.

c. After X- or γ-ray irradiation, on the curves of thermal luminescence of heated crystals appear high-temperature peaks in the area of 450–500 °C, the intensity of which grows with an increase in the temperature of heating synchronously with the growth of intensity of the ~860 nm band in the spectrum of X-ray luminescence.

d. The character of the kinetics of build-up of luminescence with a constant excitation is identical for the ~285, 480, and 710 nm bands and directly inverse for the infrared bands (Fig. 4.19). Time of establishment of dynamic equilibrium changes from sample to sample and consists of approximately 5–10 min.

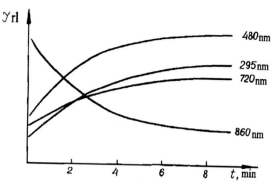

FIGURE 4.19
Curves of the build-up of X-ray luminescence of crystals of amazonites in various bands of luminescence at room temperature.

The study of spectra of X-ray luminescence of potassium feldspars has indicated the non-elementary structure of the ultraviolet band of luminescence: it consists by least measure of two bands, the relative intensity of which is practically impossible to determine due to their substantial overlap. In many cases the non-elementarity of the 285 nm band can be judged with certainty in terms of the replacement of the maximum or its broadening. In amazonites, in distinction from other types of potassium feldspars, similar broadening is observed rather frequently, and in the spectra of many samples can be distinctly identified two maximums at 280 and 296 nm. This phenomenon can be explained in various ways. First, it can be proposed that both bands are conditional upon various modifications of the same activation center (and different local coordination, symmetry, etc.); second, the bands can belong to different activation centers; and finally, the bands can be determined by centers in particular defects of the lattice, the composition of which as *disturbants* includes various impurity zones.

The first sufficiently substantiated model of the centers of ultraviolet luminescence in amazonite was proposed by A.N. Tarashchan with collaborators in 1973. They linked radiation in the ~285 nm band with the electron transition $^3P_1 \rightarrow {}^1S_0$ in ions of Pb^{2+} that replace K^+. This allows an explanation for a range of experimental facts:

1. The coincidence of the spectral situation of absorption bands of ions in feldspar ($\lambda_{max} = 255$ nm, transition $^1S_0 \rightarrow {}^3P_1$) and excitation of the ultraviolet band of luminescence
2. The dependence of the intensity of the 285 nm band on the concentration of impurity lead in crystals of amazonite
3. The parallel increase in intensity of the 255 nm absorption bands and the photoluminescence of the ~285 nm bands with thermal decoloration of crystals of amazonite (having been heated to temperatures of 550–600 °C), contingent upon an elevation of concentration of optically active centers of Pb^{2+} in consequence of thermal ionization of centers of Pb^+: $Pb^+ - e \rightarrow Pb^{2+}$

4. A decrease of activation absorption with X-ray irradiation of crystals of amazonite—the so-called effect of negative induced absorption, which is contingent upon an altering of the charge of impurity absorbing an ion with a higher valence than that of the replacing cation of lattice; this has allowed the conclusion that in natural amazonites exist two types of centers of lead: stable Pb^{2+} (not altering the charge with X-ray or γ-irradiation), associated with structural or impurity defects, and unstable (high-symmetry) Pb^{2+}, which easily alters with thermoradiational treatment its charge ($Pb^{2+} \Leftrightarrow \gamma$-irradiation or $T \rightarrow Pb^+$) and the predetermined formation of hole centers of O^- in one of the oxygen ions.

According to the results of the spectral luminescence research of M.N. Ostrooumov, relative concentrations have been calculated of irradiation-induced (or occurring in natural samples) Pb^+ centers. G.V. Kuznetsov and collaborators in 1983 proved through experiments that Tl^+ ions are effective activation centers of ultraviolet luminescence (band with $\lambda_{max} = 290\,nm$) in potassium feldspars. T.I. Sidorovskaia and collaborators proposed another model of the center: an impurity-vacancy dipole of $Me^{2+} - Vk$, in which the role of Me^{2+} is played by isomorphic impurities of Pb^{2+}, Cu^{2+}, and others.

Accounting for the numerous literature data on the paramagnetic centers (Table 4.14) and luminescence of oxygen-containing natural and artificial compounds, it follows to apply certain correctives to the existing models of centers and to examine the process of luminescence in the 285, 480, and 710 nm bands as an electron recombination luminescence that arises, in the opinion of Ch.B.

Table 4.14 Principal Paramagnetic Centers in Amazonite

Type of center	g_{xx}	g_{yy}	g_{zz}	Characteristics
Al–O⁻–Al	2.0043	2.0070	2.0555	Observed only at 77 K, thermally stable
	2.0036	2.0067	2.0275	
Si–O⁻...–x	2.004	2.0098	2.0123	Stabilized by divalent cation in the T-position (Mg^{2+}, Be^{2+})
Si–O⁻–Si	2.0046	–	2.0599	Without superfine structure, detected only in amazonites where
Pb–O⁻...–x	1.9883	2.0257	2.0561	X = Mg^{2+} or Be^{2+} in the T-position
Pb–O⁻	1.989	2.0247	2.0579	At 280 °C disrupted over the course of 1.5 h
Pb⁺	1.390	1.565	1.8369	Observed only in amazonites, thermally stable. 450–500 °C over the course of 1.5 h
Al–O⁻	2.0505	1.9911	2.0225	Stable up to 300 °C
E1 — center $(SiO_3)^{3-}$	2.0059	2.0033	2.0081	Stable up to 300 °C
Fe^{3+}_{Al}	2.00	2.00	2.00	No data

N.B. The table is comprised according to data of the author et al. [47], as well as of A.S. Marfunin, L.V. Bershov, I.V. Matiash, B. Speit, I. Lehman et al.

Lushchik, with the subsequent capture of centers of luminescence of a hole and subsequently of an electron.

Thus, it can be concluded that, independently from the details of the mechanism of the examined processes, the final stage of the excitation of luminescence of amazonite in the 285, 480, and 710 nm bands (X-ray activation) is the recombination of electrons in the hole centers of type O^- (transition $^3P_1 \rightarrow {}^1S_0$). In light of these understandings, it follows that, as the most accepted model of the 285 nm center in amazonite, explaining most fully all the experimental data at hand, we consider the complex center $Pb^{2+}-V_c-O^-$ (in other words, the center O^-, stabilized by the impurity-vacancy dipole $Pb^{2+}-V_o$). The high thermal stability of this center ensures its electrical neutrality and relatively low migrational stability of cationic vacancies. Henceforth, for the sake of simplicity, this center will be designated as Pb^{2+}. A similar model of a center of ultraviolet luminescence (Pb–O–...x) has been proposed by B. Speit and I. Lehman [49].

The wide band in the dark-blue/light-blue part of the spectrum is linked directly with the radiation of O^{2-} ions (electron transition $^3P_1 \rightarrow {}^1S_0$), included in the composition of the complex centers of type $Al-O-Me^{3+}$, the most widespread of which for potassium feldspars is, according to the data of A.S. Marfunin, the center Al–O–Al. The non-elementarity of this band, its significant broadening, and the replacement of the maximum, the value of which markedly changes for different samples and still more strongly with the transition from some feldspars to others, are explained by the complex set of the centers of luminosity close in energy parameters, distinguished only by the character of local compensation of O^- ions. The author et al. [47] suggest that the most short-wave luminescence (with $\lambda_{max} = 380-420$ nm band) is caused by centers of type $Si-O^- - \cdots V_o^{2+}$ (the concentration of which, as the experiment shows, grows with the duration and temperature of heating) and by the centers of type $Si-O^- - \cdots V_c^-$.

In the case of the centers $Fe-O^--Al$ (or $Fe-O^--Si$) emission with X-ray excitation occurs with intra-center d-d-transitions in the Fe^{3+} ions. However, in distinction from the aforementioned centers of luminescence of tetrahedrically coordinated ions, Fe_{IV}^{3+} is well excited also by the optical method. The spectrum of excitation of Fe_{IV}^{3+} ions in amazonite (see Fig. 5.3 below) consists of a range of bands of various intensity and half-width, corresponding to the spin-forbidden transitions from the principal state $^6A_1$ (S) in the above-described excited quartet states.

The identification of excitation bands of an Fe_{IV}^{3+} ion (Table 4.15) shows that the observed splitting of therms into the maximum possible number of components is contingent upon a low symmetry of the crystalline field acting on the Fe^{3+} ion. The maximum of the band of radiation is located in the 710 nm range and is weakly dependent upon the impurity composition of the amazonite. The long-wave replacement of the maximum is connected with the mixture of the albitic phase (the maximum of the band of radiation of Fe^{3+} in albite is located in the 740–750 nm range).

The identical character of the temperature dependence of the intensity of photo- and X-ray luminescence of Fe_{IV}^{3+} ions in amazonite (see Fig. 4.18(a)) provides

Table 4.15	Position and Identification of Bands in the Spectrum of Excitation of Luminescence of Fe^{3+}_{IV} ions in amazonite (77 K)	
	Position of band	
Transition	**λ_{max} (nm)**	**ν (cm^{-1})**
$^6A_1(^6S) \rightarrow {}^4T_1(^4P)$	346	28,900
	365	27,400
$^6A_1(^6S) \rightarrow {}^4E(^4D)$	371	26,594
	379	26,385
$^6A_1(^6S) \rightarrow {}^4T_2(^4D)$	394	25,380
	413	24,213
$^6A_1(^6S) \rightarrow {}^4E \; {}^4A_1(^4G)$	424	23,585
	430	23,256
	438	22,831
$^6A_1(^6S) \rightarrow {}^4T_2(^4G)$	472	21,186
	495	20,202
	506	19,763
$^6A_1(^6S) \rightarrow {}^4T_1(^4G)$	560	17,857
	610	16,393
	642	15,576

evidence that the fall of intensity of luminescence with a growth of temperature is contingent upon intra-center quenching, in consequence of which in the spectra of high-temperature peaks of thermal luminescence (>200 °C) of Fe^{3+}_{IV} bands are practically absent, and the recombination of electrons in the hole centers of O$^-$–Fe occurs without emission. Not dwelling on the details of the mechanisms of energy transfer and processes of quenching, we note only that the Fe ions fundamentally influence the character of radiation of other impurity and "lattice" centers in amazonite. Analysis of the spectra of a great quantity of samples leads to the conclusion that in the curves representing the correlations J_{Pb}–J_{Fe} and J_{O^-}–J_{Fe}, can be delineated two sections—the range where J_{Pb}–J_{O^-}—is not dependent upon the content of iron in the samples of amazonite (the range of low intensity of the 710 nm band, Fig. 4.20), and the range where an inverse correlation is observed between the intensity of bands of emission of centers J_{Pb} (J_{O^-}) and J_{Fe} (the range of intense quenching, contingent upon Fe^{3+}_{IV} ions). Therefore, with the elevation of Fe^{3+}_{IV} concentration in the crystals of amazonite is practically always noted a fall in intensity of the 285 and 480 nm bands.

The centers of luminescence responsible for the 880 nm band of luminescence in crystals of amazonite remain practically unstudied. On the one hand, this is explained by the fact that the given radiation located in the infrared band has been discovered relatively recently, and on the other, by the fact that the corresponding analogues in the activated synthetic crystals of potassium feldspar are absent. The first proposals on a connection of the 880 nm band with Fe ions (I.I. Matrosov et al.) have not subsequently received experimental confirmation—880 nm centers occur in sharp distinction from Fe—centers of

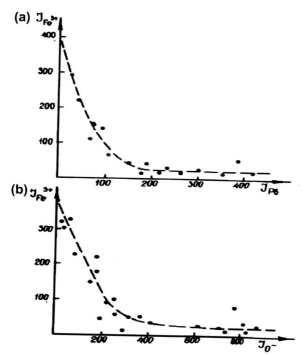

FIGURE 4.20
Correlation of intensities of luminescence of J centers Pb^{2+} and Fe^{3+} (a), as well as O_{Al}^- and Fe^{3+} (b) in amazonites from various deposits.

spectra of excitation. In recent time, interest toward these centers has fundamentally increased, insofar as they are intrinsic only to potassium feldspars and most characteristic for microclines of rare-metal pegmatites.

With the research of infrared luminescence of amazonite has been established the following:

1. In natural samples of amazonite, luminescence in the 860 nm band arises with photo- as much with X-ray excitation.
2. In the spectrum of excitation are observed two bands with maximum of 380 nm (3.25 eV) and 550 nm (2.25 eV).
3. With photo- and X-ray excitation, the intensity of infrared band depends upon the temperature of heating of the sample; furthermore, short-term heating up to 450–500 °C in air and extended heating (6 h) in autoclave in water steam (T = 400 °C; P = 100 Mpa) practically does not alter the intensity of the band; and at temperature > 650 °C, the infrared photoluminescence disappears, while the intensity of X-ray luminescence grows.
4. In annealed samples of amazonite, infrared luminescence appears already with short-term X-ray irradiation, and its intensity grows in parallel with the growth of the high-temperature peak of thermal luminescence with T_{max} = 470–490 °C.

At the moment of activation of the X-ray radiation, an infrared flare is noted in the 860 nm band, the intensity of which decreases during several seconds. The initial intensity of the flare depends upon the temperature and duration of annealing.

All of these facts accounting for the character of the kinetics of the build-up of X-ray luminescence in artificial crystals (see Fig. 4.19) provide the basis for concluding that in the crystals of amazonite, infrared luminescence is contingent upon electron centers, most probably the centers in Pb^+ ions. Infrared luminescence of Pb^+ ions was detected by L.E. Nagli and S.V. D'iachenko in 1986 in crystals of KCl–Pb irradiated by X-rays at 300 K. With X-ray excitation of amazonite, luminescence is the result of the dissociation of the exciton on isolated Pb^{2+} ions (capture or localization of the electron) with the formation of excited Pb^+ centers, the transition of which to the normal (basic) state occurs with the obtaining of light quanta with long 860 nm waves (1.44 eV). Schematically, this process can be described as follows:

$$Pb^{2+} + e^o \rightarrow Pb^{2+}e^- + e^+ \rightarrow (Pb^+) + e^+ \rightarrow Pb^+ h\nu_{860} + \cdots + O_{Al}^-,$$

where e^o, e^- and e^+ are respectively exciton, electron, and hole. The proposed model of the infrared center satisfactorily explains the experimental data in terms of a change of the intensity of the 860 nm band as the result of a high-temperature (greater than 700 °C) annealing of crystals of amazonite, which leads to the disruption of complex centers of amazonitic color and an increase in concentration of isolated Pb ions. This process is analogous to the natural process of the breakdown of amazonitic color (de-amazonitization), with which also occurs an increase in the intensity of the 860 nm band. As indicated by the experimental data, the intensity of the bands of X-ray luminescence of crystals with flaws of unevenly distributed color is varied. Furthermore, independently from the degree and direction of the change of intensity of other bands (285, 480, and 710 nm), for the 860 nm band is always observed an (occasionally significant) increase in intensity with the transition from the transparent isolated sections with amazonitic color to sections of white, yellowish, and other non-amazonitic color (Fig. 4.21). It is likewise not improbable that F^+ centers can be centers of infrared luminescence in amazonite (i.e., a double-charged vacancy of oxygen with one captured electron $V^{2+} e^-$).

Insofar as oxygen vacancies, as are Pb^{2+} ions, are electron traps, the mechanism of the processes of luminescence of F^+ centers with X-ray excitation should be analogous to the mechanism of luminescence of Pb^+ centers.

The principal types of centers of luminescence observed in amazonites and their spectroscopic characteristics are cited in Table 4.16.

The spectral luminescence characteristics of amazonites from concrete pegmatite and granite deposits supply the basis for the following conclusions. In the spectra of amazonites from Precambrian pegmatites of the alkaline-granite formation, the primary role is played by Pb^{2+} and Al–O$^-$–Al centers, whereas in spectra of

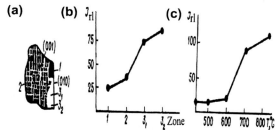

FIGURE 4.21
Distribution of color in various sections of an individual of microcline (a): 1–3—color (1: green, 2: pale-green, 3_1: yellowish, 3_2: brownish); change in intensity of the 860 nm band J_{rl} in variously colored sections (b), and with isothermal annealing of green amazonite at various temperatures T (c).

Table 4.16	Principal Types and Characteristics of Centers of X-ray Luminescence in Natural Amazonite		
Type of center	**Model**	**Bands of radiation, λ (nm)**	**Half-width of band of radiation (eV)**
O_{Al}^-	Al–O$^-$–Al	480	0.7
$O_{V_c}^-$	V_c^-–O$^-$–Si	380–420	0.6
	V_c–O$^-$–Si		
Pb^{2+}	Pb^{2+}–V_c^-–O$^-$	285	0.4
Fe_{IV}^{3+}	Fe_{Al}^{3+}	710	0.2
Pb^+	Pb_K^+	860	0.2

amazonites from relatively younger and less deep pegmatites, the preeminent position is occupied by Fe^{3+} and Pb^+ centers. The latter are generally typical for amazonites from pegmatite deposits with rare-metal mineralization. The same can be said about the spectra of amazonites from rare-metal granites in which, along with infrared centers, are fairly widely distributed ultraviolet centers and Al–O$^-$–Al centers, whereas Fe^{3+} centers are manifested very weakly. In practically oreless granites are noted amazonites in the spectra of which predominate only Fe^{3+} centers; the rest play only an insignificant role. Thus, the particularities of the spectra of luminescence of amazonites can serve as a fairly prospective indicator of the development in amazonite-containing rocks of rare-metal mineralization.

It is interesting to trace the changes in the intensity in bands of emission of various centers and their concentrated correlations in amazonites of various sections of pegmatite bodies and on the whole in all studied deposits. First and foremost, it is necessary to mention that the change of X-ray luminescence parameters of amazonite is closely connected with the change of its color in all zones of the studied pegmatite bodies. Moreover, in the distribution of the intensity of separate bands of emission (in the general case of proportional content in crystals of

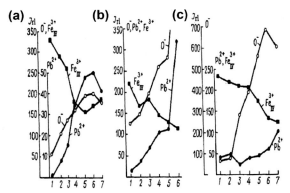

FIGURE 4.22

Character of the change in intensities of bands in the spectra of X-ray luminescence J_{rl} of potassium feldspar from various zones of pegmatite bodies. (a and b) Il'menskie Mountains, Urals (a: mine 70, b: mine 57); 1–3: microcline (1: coarse-crystalline pink from the peripheral zone, 2: pink at 30 cm from the contact, 3: greenish from the fine-graphic zone); 4–7: amazonite (4: green from the graphic zone, 5 and 7: bluish-green from the coarse-graphic zone, 6: blue from the central part of the vein); (c) Kola Peninsula, Ploskaia Mountain; 1–7: amazonite (1 and 2: from the pegmatoid zone (1: blue, 2: green); 3–5: from the fine-blocky zone (3 and 4: bluish-green, 5: grass-green), 6 and 7: dark-green from the blocky zone).

corresponding optically active centers) is revealed a fairly distinct zonality that is symmetrical with regard to the center of the vein. With the transition from the large-crystalline pink microcline of the peripheral zones to the greenish-blue amazonite from the central part of the veins (Il'menskie Mountains, mines 70, 57) occurs a regular decrease in the intensity of the luminescence of Fe^{3+} ions, and an increase in the intensity of the bands of the oxygen centers (Fig. 4.22(a) and (b)). The same is observed for the pegmatite body with the zonality distribution of the color amazonite generation (Fig. 4.22(c)): the intensity of the luminescence of Fe decreases and the intensity of oxygen and lead Pb^{2+} centers increase with a transition from blue amazonites of the peripheral zones to the green potassium feldspars of the central part of the body. This transition correlates with the general concentration of lead and iron in these crystals from 0.1% Pb and 0.08% Fe in the blue peripheral amazonites to 1% Pb and 0.03% Fe in the green samples from the central part of the pegmatite vein.

Such regularity in the change of intensity of the bands of luminescence of the activation and lattice centers in amazonites is characteristic not only for separate pegmatite bodies, but also for deposits on the whole (Fig. 4.23; see Fig. 4.22). This can be explained above all by the particularities of distribution and by the concentration of optically active centers in potassium feldspars, which influence in a fundamental way the spectral characteristics of luminescence and the particularities of the recombination processes. In part, with notable concentration impurity centers are clustered, and the spectrum of activated crystals is determined by exchange coupling (Pb–Fe and the like). On the other hand, with the growth of concentration of Fe_{IV}^{3+} centers, a fall of intensity of bands of O^- centers occurs as much in consequence of a decrease in their concentration as

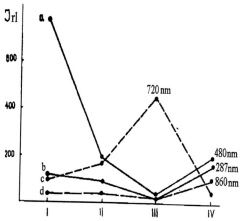

FIGURE 4.23

Change in intensity of bands in the spectra of X-ray luminescence of amazonite J_{rl} from various deposits: (I) Kola Peninsula, (II) Urals, (III) Kazakhstan, (IV) Transbaikal; (a) $\lambda_{max} = 480$ nm (O_{Al}^-); (b) $\lambda_{max} = 287$ nm (Pb^{2+}); (c) $\lambda_{max} = 720$ nm (Fe^{3+}); (d) $\lambda_{max} = 860$ nm (Pb^+).

of the processes of the external quenching of luminescence, which can explain the competition for capturing free charge carriers in recombination processes between the $O_{Al, Si}^-$ and O_{Fe}^- centers.

The UV-emission spectra of the green and the bluish-green untreated amazonites are quite similar, but they differ considerably from that of the blue amazonites [47]. Each of the samples shows three UV-emission maxima at or about 465, 678, and 741 nm. Comparison of the UV-fluorescence spectra of the untreated samples with those of the treated ones reveals changes of fluorescence caused by irradiation. As mentioned above, the UV-emission spectra of the green and bluish-green untreated amazonites are similar. Accordingly, their emissions spectra after the X-radiation are also similar, i.e., the main change is observed between 360 and 469 nm. Ostrooumov et al. [1] reported that these bands correspond to the concentration of Si–O−–Si and Al–O−–Al hole centers, the concentration of which increases after irradiation. It should be noted that in contrast to the green and bluish-green samples, the change caused by irradiation in the case of the blue amazonite occurs mainly at 680 nm and 741 nm; that is, the intensities of these two bands are increased, whereas the region between 362 and 460 nm remains more or less unchanged. The bands at approximately 700 nm are characterized by the typical spectrum of $^{IV}Fe^{3+}$, in which early blue generation-I amazonites with a maximum concentration of Fe (see Table 4.1) are included.

4.4.4   Vibrational Spectrometry of Amazonite

4.4.4.1   INFRARED SPECTROMETRY

One of the important questions connected with the crystal chemistry of amazonite is the presence in its structure of isolated hydroxyl groups as much as

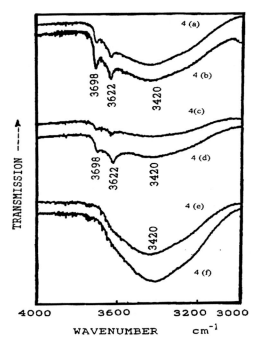

FIGURE 4.24
IR transmission spectra of amazonite showing changes of the OH-stretching vibrations before
(b, d, and f) and after (a, c, and e) irradiation: (a, b) = green amazonite; (c, d) = bluish-green amazonite;
and (e, f) = blue amazonite.

of molecular water, which in correlation with a range of known models participates in the formation of centers of color [28,34,46]. In this case, necessary information can be obtained from the infrared spectra of this mineral.

The IR transmission spectra in Fig. 4.24(a–f) show the changes of the OH-vibrations in the amazonite samples before and after irradiation. As may be observed, all of the samples contained structural H_2O. The green and the bluish-green amazonites show two distinct absorption bands at 3698 and 3622 cm^{-1} due to OH-stretching vibration, and one absorption band at 3420 cm^{-1}, which indicates the presence of H-bonded OH. The spectra of the blue amazonite, on the other hand, lack absorption bands at 3698 and 3622 cm^{-1}, but display a band at 3420 cm^{-1}, indicating that it contains only H-bonded OH.

Accordingly, the OH-stretching vibrations of amazonite also change due to irradiation. As shown in Fig. 4.24(a–f), the intensities of the absorption bands related to the OH-stretching vibrations mentioned above decreased considerably with X-ray irradiation of amazonite. Figure 4.24 shows that the intensity of most of the bands decreases in variable amounts due to irradiation.

The obtained spectra of infrared reflection of amazonites confirm a decrease in the intensity of the majority of bands following X-ray irradiation [47].

A question no less important than that of obtaining information from the infrared spectra regarding the nature of water in amazonites is that of determining their structural order. A range of research has already long proven the possibility of the evaluation of the degree of order of alkaline feldspars in terms of the infrared spectra of transmission of their prepared powder samples. An analogous evaluation, as has been indicated by M.N. Ostrooumov [17,33], can be carried out by means of the infrared spectra of reflection.

In correlation with the known data in the literature, the most perceptible absorption bands that reflect the degree of order of alkaline feldspars are the bands with 640–650 cm^{-1} (v_1) and 540–550 cm^{-1} (v_2) that change only their position, and two bands in the 700–800 cm^{-1} range (720–730 and 760–780 cm^{-1}) that are stable in terms of position but sharply variable in terms of intensity. Correspondingly, there are two methodologies of evaluating the degree of structural order of alkaline feldspars in terms of their infrared spectra of transmission.

Bands v_1 and v_2 are caused by valence symmetrical vibrations in silica tetrahedra and reflect a change of the force constant between O, Si, and Al in the tetrahedra of type T_2 (v_1) and tetrahedra of type T_1 (v_2). The value of $\Delta = v_1 - v_2$ is a characteristic of the degree of order of alkaline feldspars, varying from 90 cm^{-1} in high sanidine to 100 cm^{-1} in maximum ordered microcline. The corresponding value of infrared order is calculated in terms of the empirical formula $\Theta = 0.05$ ($\Delta v - 90$).

On account the aforementioned, it follows to note that all amazonites from Precambrian pegmatites in terms of value of Θ belong to the maximum ordered microcline. At the same time, it follows to note the low structural disorder of the later generations of low-temperature amazonites.

In amazonitic granites are commonly identified several generations of alkaline feldspars, which include amazonites. Furthermore, amazonites of early generations are distinguished from those of later generations by lower values of infrared order in the 0.40–0.75 interval that corresponds to the disordered varieties of orthoclase. On the whole, the infrared spectra confirm all of the results that were obtained with the aid of X-ray diffraction methods. With the aid of correlative analysis, an evaluation has been conducted of the strength of the connection of the X-ray-graphic and infrared-spectrometric characteristics of the structural state of amazonites. As a result, it has been established that the greatest strength of connection, approaching the functional, is noted between the ratio I_{730}/I_{770} and the degree of monoclinic order (Δt).

Analogous results have been observed also in terms of the infrared spectra of reflection of various generations of this mineral in pegmatites and granites. From our point of view, infrared reflection spectrometry by comparison with X-ray methods is a fairly universal, simple, and express method of evaluating the degree of structural order of amazonites and alkaline feldspars. There is a range of advantages accruing to this methodological variant also compared to the infrared spectrometry of transmission. In part, with its aid can be removed the following fundamental limit, particular to the infrared spectra of transmission—the possible superposition on the absorption bands of amazonites of characteristic bands of other minerals located with them in close intergrowth of the mineral phases manifested in the same ranges of the spectrum. On this account, with attention directed as well to the significant growth in topomineralogical research that call for study of a greater sample quantity, it is proposed to examine the method of infrared reflection spectrometry in the capacity of a simple and informative variant of evaluating the degree of Si-Al order of amazonitic feldspar [36].

4.4.4.2 RAMAN SPECTROMETRY

The method of Raman spectrometry allows in a range of cases to fundamentally define with more precision the data obtained by X-ray-structural analysis as well as by infrared spectrometry [45]. It is particularly valuable with the study of light atoms, their isotopes and groupings, in particular the hydroxyl group [42]. At present, this method occupies the most important place in the study of natural minerals, which as a rule are characterized by the presence of impurities, disorder of lattice, superstructure, and structural decay [30,38,39,44].

Objects of research have been the samples of amazonites from Precambrian and Paleozoic pegmatites as well as Mesozoic granites (Kola Peninsula, Urals, Transbaikal). The Raman spectra were obtained with the aid of spectrometers Jobin Yvon 64,000 and Bruker RFS100 with excitation of the 514 nm lines by Ar$^+$ laser and 1064 nm Nd/YAG radiation. The utilization of various sources of excitation has enabled a reduction in the various experimental errors connected with the heterogeneity of the researched crystals.

In the first order, in the Raman spectra of amazonites from the range of studied deposits, it follows to focus attention on the thick band in the 2700–3200 cm^{-1} range, which as a rule is not manifested in spectra of other feldspathic species and varieties (Fig. 4.25). With attentive examination, it can be established that this band is characterized by a fairly complex structure, that is, the identification within it of a series of lines of corresponding vibrations of molecular water and hydroxyl groups. It follows to note that the greatest intensity of this band is particular to the Raman spectra of Precambrian amazonites, at the same time as in Raman spectra of amazonites from Mesozoic granites are reported bands in this short-wave range minimal in terms of intensity.

Moreover, in the obtained Raman spectra of amazonites is observed a great number of bands (Table 4.17) corresponding to lattice vibrations, particularly in the interval of frequencies from 500 to 550 cm^{-1}. The width of these bands as a rule is not great, while the intensities of several of them are fairly significant. Furthermore, changes of intensities of certain of the bands of this interval depending on survey conditions allow them to be classified to the vibrations of Ag or Bg in the case of utilization of the monoclinic approaching. In the range of higher frequencies, Raman bands of amazonites are fundamentally less intensive—evidently characteristic for structures with framework silicate lattice

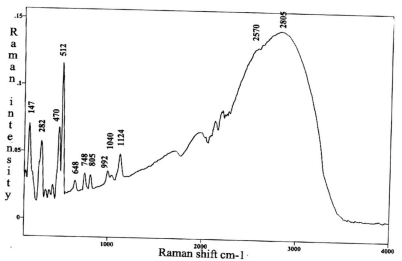

FIGURE 4.25
Raman spectrum of green amazonite from pegmatites of the alkaline-granite formation.

Table 4.17   Frequencies of Bands in the Raman Spectrum of Amazonites (cm⁻¹)

Amazonite	Sanidine		Amazonite	Sanidine	
	Ag	Bg		Ag	Bg
1136 (Ag)	1130	–	370 (Ag)	379	–
1123 (Ag)	1125	–	285 (Ag)	286	–
1120 (Bg)	–	1120	257 (Bg)	–	260
1095 (Bg)	–	1120	200 (Bg)	–	196
993 (Bg)	–	973	180 (Bg)	–	170
812 (Ag)	810	–	158 (Ag)	162	–
765 (Ag)	784	–	141 (Ag)	144	–
650 (Bg)	–	650	127 (Ag)	130	–
630 (Ag)	638	–	110 (Bg)	–	101
580 (Bg)	–	578	95 (Ag)	97	–
512 (Ag)	515	–	82 (Ag)	81	–
475 (Ag)	476	–			

and their related compounds. These bands correspond to valence vibrations of the bond Si–O–Si and Si–O–Al, distinguished by widths that are great in terms of intensity and change very significantly depending on experimental conditions, which complicates their description by vibrations, similar to the vibrations of types Ag and Bg of the monoclinic crystal. In correspondence with the obtained data, amazonites from various genetic and age types of granitoid rocks are practically indistinguishable in terms of their Raman spectra in the range of lattice and valence vibrations.

The Raman method, as has already been noted, is widely accepted for the study of the disordering of the crystalline lattice of minerals. The broadening of the vibrational lines, connected with the distortion of order, is distinctly manifest in Raman spectra, the lines of which are in terms of their nature narrower than the lines of infrared absorption [33]. Thus, in natural albite that has undergone an extended stage of lowering of temperature, the distribution of Al and Si atoms is ordered in terms of the crystallographically nonequivalent tetrahedral positions T_1 and T_2. Its Raman spectrum is represented by a series of narrow lines. In the same albite that is subjected to extended heating and rapid cooling, X-ray data reveals a strong disorder, in which is distinctly correlated the broadening of all lines in the Raman spectrum. The distortion of the periodicity of the most aluminosilicate structure is characterized by an extended flattening of the spectrum, the disappearance in the first order of the low-frequency lines, and the maintenance of the wide lines correlated with the internal vibrations of the silica groups [29,44].

It is known [17,33] that the polymorphic modifications of potassium feldspars are distinguished by the distribution of Si and Al along 16 tetrahedral positions of the unit cell, remaining practically unchanged for all modifications. Its composition can be recorded as $4xKAlSi_3O_8$ or $4xKT_4O_8$ in case of the full statistical distribution of Si and Al. The last case is realized only in high sanidine with symmetry of C2/m: all tetrahedra TO_4 are identical in terms of size and occupy the two crystallographic positions $8T_1$ and $8T_2$ with averaged composition $Al_{1/4}$ and $Si_{3/4}$ each. A cell consists of 32 atoms of oxygen, distributed according to the following set: $4O_{A1}$ on the axes C2; and $4O_{A2}$ in the σ_h and $8O_B$, $8O_C$, $8O_D$ plates, in the general positions. The environment of the atom T_1 is formed by O_{A1}, O_B, O_C, and O_D and of the atom T_2 by O_{A2}, O_B, and O_D; the average distances of T_1–O and T_2–O are identical and equal to 1.642 Å.

A more significant order of distribution of Al/Si with maintenance of symmetry of C2/m is intrinsic to the orthoclases, and the greatest, though varying for samples of different origin, degree of order is reached in microcline with lowered symmetry of lattice up to triclinic (C1). In consequence of the given lowering of symmetry, each set of a ratio of eight breaks down into two ratios of four. For example, $8T_1$ corresponds in microcline to $4T_1(0)$ and $4T_1(m)$, with an analogous breakdown for the set $8T_2$. The average lengths of connections of T–O fundamentally differ as they are connected with different contents of aluminum in tetrahedra of a different type, which, however, in principle is concentrated in tetrahedra $T_1(0)$.

The situation described above signifies that with the interpretation of vibrational spectra of the aluminosilicates structure of potassium feldspars, in part of amazonite, in the first order it follows to consider the difference in the dynamic properties of bonds from tetrahedra T_1 and T_2. The data in the literature on Raman spectra of potassium feldspars are exhausted in several researches containing information on the spectra of sanidine and microcline with attempts at identifying the frequencies corresponding to fundamental vibrations of crystals

[5]. In result, a majority of the fundamental vibrations of microcline symmetrical to the inverse have been identified, and their correspondence of irreducible representations of the group of the symmetry of the approaching monoclinic model of lattice of this crystal has been established.

In result of the study of Raman of amazonites in all stages of heating (within various ranges of temperatures of thermal discoloration up to 1000 °C), it has been established that the spectral characteristic parameters provide evidence of the stability of the structural state of amazonitic potassium feldspars before and after high-temperature heating. In other words, in terms of the Raman spectra, in this case, a process of structural Si-Al-disordering of amazonites is not reported, nor is their transition to the monoclinic modification. Thus, these spectrometrical data support the results of the full X-ray-structural analysis (see Section 4.3), in correspondence with which, likewise, has not been detected the phenomenon of disordering of amazonites after their multi-hour medium- (300–500 °C) and high-temperature (900–1000 °C) annealing.

4.4.5 Electron Paramagnetic Resonance

It is known that electron paramagnetic resonance (EPR) spectra can be observed and analyzed in all matter and in any aggregate states if they contain elements with unpaired spins [34]. To these, in part, belong electron–hole centers, that is, electrons or holes localized in the defect positions of a crystalline structure.

Marfunin[30] first reported EPR spectra of amazonite taken at 9.3 GHz and 78 K, and the presence of a range of paramagnetic centers in this alkaline feldspar was first indicated. An electron center ascribed to Pb^+ was observed only in amazonites. The spectrum consisted of a central line with two hyper-fine structure lines, which can be described with a spin Hamiltonian with orthorhombic symmetry and a g-factor of 1.390, 1.565, and 1.837. The amazonite was heated to 400–500 °C for several hours until it lost its characteristic color and Pb^+ spectrum. The EPR spectrum ascribed to Fe^{3+} was unchanged by heating. These authors concluded that the color of amazonite results from Pb^+ produced by the substitution of Pb^{2+} for K followed by the capture of an electron.

The change of color and the formation of defect centers in different feldspars (including one amazonite) after X-ray irradiation have been investigated using EPR spectroscopy by Speit et al. [49] and by Hofmeister and Rossman [28]. Recently, using EPR, two nonequivalent Pb ions in an amazonite-type microcline structure were indicated [48]. The problem is that in these publications there has been no explanation for the difference between EPR signal and the color of the principal generations of amazonite from different granitoid formations, that is, the presence in nature of a range of colored varieties of amazonite. This once more confirms the necessity of a differentiated approach to the study of amazonites from various genetic types of granitoid rocks.

We indicate here only the difference in EPR spectra of amazonites from two genetic types: Precambrian REE pegmatites of the Kola Peninsula and Mesozoic

Table 4.18	EPR Data for the Pb Centers in Amazonite				
Center	g_z	g_y	g_x	Interpretation	Characteristics
I	1.390	1.565	1.837	Pb^+ electron center	The EPR signals of Pb^+ and the absorption band at 625 nm disappear after heating to 450–500 °C.
II	1.989	2.0247	2.0579	O^-–Pb hole center	The band at 740 nm and the EPR signals of O^-–Pb hole center disappear in a heating process of up to 280 °C for 1–1.5 h.

granites of Transbaikal. In the former by the EPR method have been studied two principal generations: early blue and late green amazonites.

EPR spectra of the amazonite varieties show some characteristic lines in the two zones between 5000 and 3500 Gauss and 3400–3200 Gauss. The same peaks appear systematically in the spectrum of the studied samples. The associated paramagnetic centers are shown in Table 4.18. Starting from principal values of the g-factor at the center I allowed to attribute the peaks to a Pb^+ electron center as described above. Center II was interpreted as a result of an O^-–Pb hole type center by Petrov et al. [48] and Speit and Lehman [49]. The axis of the g-tensor of these two centers is in parallel. The Center II disappeared in a heating process of up to 280 °C for 1–1.5 h, whereas Center I remains stable at these temperatures. Resonance attributed to a Pb^+ electron center was most common in the blue variety of amazonite (early generation), whereas an $O–Pb^+$ hole center is more typical for the green amazonite (late generation). The room temperature spectrum of natural bluish-green resulted from the combination of these two centers.

The EPR spectra of amazonites from Mesozoic granites of Transbaikal are characterized by the presence of only one type of defect center: Pb^+, the electron center of single-valence lead. Thus, a marked difference is established between early and late generations of amazonite in terms of the type of electron–hole centers, and a typomorphic significance of EPR spectra is noted for amazonites from various genetic types of granitoid rocks.

Furthermore, it follows to focus attention on the fact that in EPR spectra of all colored varieties of amazonite is always reported the presence of hole center $Al–O^-–Al$ ($g_z = 2.0043$, $g_y = 2.0070$, $g_x = 2.0555$), which, as is known, is the most widespread in the various feldspars. With the entry of Al^{3+} into the position of Si^{4+}, the arising surplus of negative charge leads to the output of the electron in this defect tetrahedra $Al–O_4$. The deficit of a negatively charged electron, described as a positive hole, is localized predominantly in the oxygen ion, which interacts with two ions of aluminum: one in the normal tetrahedron, and the other in the defect (impurity)—$Al–O–Al$. The noted center is more concentrated in green amazonite.

It is known that upon irradiation with X-rays, almost all feldspars showed this type of center. That is why the irradiated green amazonite shows EPR spectra with a higher concentration of Pb and Al hole centers in comparison to blue and non-irradiated varieties. In addition to this, particularly with the green sample, it was also observed, as previously by Speit and Lehman [49], that they attain a brownish hue (formation of the Al hole centers).

The colors of all of the treated samples, at first glance, were intensified by irradiation. However, it is important to note that after X-ray irradiation, not only were the intensities of the bands at 740 and 380 nm seen to increase, but the intensity of the band at 625 nm decreases. With the heating of the green amazonite sample to 280 °C, the band at 740 nm and the EPR signals of the Center II disappear, to be replaced by the lines of Center I. Heating for several hours at 280 °C does not cause this center to disappear. The EPR signals of center I and the absorption band at 625 nm disappear after heating to 450–500 °C. Thus, the EPR signal behavior of these centers (I and II) after heating and irradiation can be well correlated with the behavior of the amazonite absorption bands at 625 and 740 nm (Table 4.18). These data do not agree with the results obtained by a theoretical study of the absorption spectra of Pb^+ and Pb^{3+} in the K^+ site of microcline [29,48].

Thus, the EPR spectra of amazonites enable us to obtain additional information on the types of electron–hole centers in this mineral and to define more precisely their interrelation with the established bands of optical absorption.

CHAPTER 5

Color and Genesis of Amazonite

5.1 MODELS OF COLOR CENTERS

Prior to the examination of potential reasons behind amazonitic color, it is necessary to characterize certain dependencies of absorption bands in the 600–750 nm region on the composition of the studied samples.

As has already been noted by a range of researchers, an almost linear dependence has been established of the intensity of the amazonitic band on the content in samples of lead impurity. Insofar as these measurements have not taken into consideration the complex structure of the amazonitic band and the readings were setting about the maximum of the envelope curve (the position of which for the spectra of various samples can vary within fairly wide limits), the author et al. [47] have repeated such experiments for the most distinctly expressed absorption band with a maximum of 630–635 nm (band II). The intensity of this band, measured in terms of the values of the optical density in the spectra of diffuse reflection (Fig. 5.1), is not directly connected with the content of lead impurity in the studied samples of amazonite. Additionally, with its concentration from 0.01% to 0.1% is revealed inverse correlation.

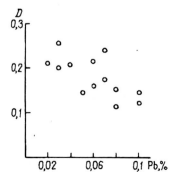

FIGURE 5.1

Dependence of the intensity of the 630 nm absorption band (D) upon the content of lead impurity in amazonites.

Amazonite: Mineralogy, Crystal Chemistry, and Typomorphism. http://dx.doi.org/10.1016/B978-0-12-803721-8.00005-6

In any event, however, the impurity of lead is not the principal reason for the origin of the 630–635 nm absorption band.

On the other hand, the elevation in the content of lead in amazonites influences, in the most fundamental way, the position of the maximum of the envelope curve (the superposition of bands II and III), shifting it in the long-wave region of the spectrum (Fig. 5.2). This fact unambiguously indicates the direct dependence of the intensity of band III (with a maximum of 720–740 nm) upon the content of lead impurity. Naturally, the increase in the intensity of band III is accompanied by the increase of the entire amazonitic band, which on the whole leads to a strengthening and change in the tone of the color of the amazonite.

Taking into account all the above-cited data that provides evidence of the participation of various chromophore centers in the coloration of amazonite, we will consider the possible nature of the absorption band of each of them.

Band II (620–650 nm) commonly dominates in the optical spectra of amazonites. It is of principal importance that the absorption band in the given region is not a characteristic particularity of amazonites: in the spectra of the latter, it possesses only an anomalously high intensity, adequate for the appearance of specific amazonitic color. As has been noted in the works of M.N. Ostrooumov, B. Speit, and I. Leman [12,49], the 650 nm (15,300 cm^{-1}) absorption band is reported in the optical spectra of many feldspars not possessing amazonitic color. A weak 650 nm band was discovered by A.N. Platonov in the spectra of the golden-yellow Madagascar ferro-orthoclase [19]. A more long-wave 720–750 nm absorption band, in the opinion of M.N. Ostrooumov and A.N. Platonov, is the reason for the pale-green color of so-called amazonite-type or amazonite-like microclines, orthoclases, and other feldspars [50].

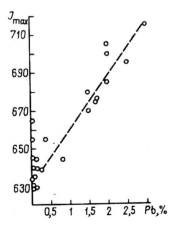

FIGURE 5.2
Dependence of total intensity of absorption bands II (625–650 nm) and III (720–740 nm) in the optical spectra of amazonite upon the content of lead impurity.

The detailed research conducted by V.T. Veremeichik and others of the optical spectra of absorption of iron-containing orthoclases [20], with the involvement of theoretical calculations of the schema of the levels of Fe^{3+} in the tetrahedral crystal field, as well as numerous measurements and interpretations of the spectra of the excitation of luminescence of various feldspars performed by A.N. Tarashchan and A.I. Bakhtin [3], have permitted the description of the wide, weak absorption band in the spectra of the studied samples (Fig. 5.3) of the spin-forbidden transition $^6A_1$ $(^6S) \rightarrow {}^4T_1$ (^4G) in the Fe^{3+} ions located in the crystal field with symmetry ^1T_d, that is, occupying the tedrahedral positions Al or Si in the structure of feldspars.

According to EPR data and luminescence research, the principal part of Fe^{3+} ions in microclines is situated in the tetrahedra T_{1o} (the largest in terms of its dimensions), which is commonly occupied by aluminum ions. A small part of the Fe^{3+} ions, depending upon the ordering of the microcline, can be situated in the more fine tetrahedra T_{1m}, T_{2o}, and T_{2m}. It has been noted as well that, in the spectra of various feldspars, the position of the maximum of the band of electron transition $^6A_1 \rightarrow {}^4T_1$ in the 600–700 nm region depends upon the dimensions of the tetrahedra enclosing the Fe^{3+} ions (the maximum is shifted in the long-wave region with an elevation of the dimensions of the tetrahedra). According to the data [3,19], in the spectra of albite and labradorite, the maximum of the band of this transition has been located within the 670–680 nm interval.

However, as has already been noted, the intensity of the examined absorption band in the optical spectra of the overwhelming majority of iron-containing feldspars is extremely low, and only for certain green amazonite-type varieties is the ~650 nm band distinguished by a somewhat elevated intensity, resulting in the greenish tones of color of these samples. It is fully probable that precisely the low intensity of this band in the spectra of Fe^{3+}-containing feldspars has served as the principal argument against its comparison with the intensive amazonitic

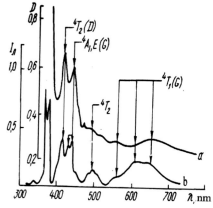

FIGURE 5.3
Spectra of optical absorption (a) and excitation of luminescence (b) of Fe^{3+} ions in microclines.

band. However, the question is warranted: which reasons or factors are most responsible that this absorption band precisely in amazonites is characterized by an anomalously high intensity?

In recent years in the field of optical spectroscopy of inorganic compounds including minerals, the most diligent attention has been devoted to the study of the particularities of optical absorption and the mechanisms of the correlation of the exchange-coupled states of ions of transitional metals. This special type of optical absorption (spectra of charge transfer or exchange-coupled pairs: metal–metal, Me–Me) plays, in the opinion of A.N. Platonov, an important role in the color of natural minerals, particularly iron-containing silicates, which include rock-forming pyroxenes, amphiboles, and micas [19]. The principal particularity of the spectra of charge transfer is the presence of a wide, intense, and sharply polarized absorption band in the 550–800 nm ($\sim$19,000–12,000 cm^{-1}) region. Such bands are typical for the optical spectra of minerals containing variously charged ions of iron (Fe^{2+} and Fe^{3+}) in neighboring structural positions, that is, in the linked general planes or edges of the structural polyhedra. In this case, the vector of absorbing light is oriented along the direction Me–Me, which results in the characteristic polarized properties of the absorption band of the Fe^{2+}–Fe^{3+} pairs.

In the literature about optical spectroscopy of minerals, this band is commonly called the type Me–Me charge transfer band, and the electron transitions responsible for it are called transitions with charge transfer. G. Smith, R. Burns, and others [30] have proposed that in the process of optical excitation of the pair of variously charged ions occurs a certain oscillation of valence on account of a partial delocalization of electrons of the donor ions (for example, Fe^{2+} ions), that is, the charge transfer $Fe^{2+} \rightarrow Fe^{3+}$ is formally realized.

On the other hand, there exists a somewhat different interpretation of similar absorption bands—metal-anion-metal (Me–L–Me) type—from the position of ionic interactions [3] occurring due to the covalent mixing of atomic orbitals of ligands and ions of metals. In this case is proposed the following schema of an exchange pair:

$$Fe^{2+} \diagdown \negthickspace^{\textstyle O}_{\textstyle O} \negthickspace \diagup Fe^{3+}$$

The two elaborated conceptions do not, in principle, contradict each other, insofar as they both originate from the proposal of a paired correlation of ions of transitional metals.

The indicated example is merely a single isolated case of what, for minerals, is actually a fairly widespread electron transition: Ti^{4+}–Fe^{2+} (dark blue sapphires and cyanites), Pb–O–Cu (greenish-red-brown cuprodescloizite, dark-green duftite), and UO_2–Pb (orange-red curite and others). In all the examined examples above is noted, according to A.N. Platonov [19], a strengthening of intensity of the forbidden absorption bands of d–d transitions. A similar mechanism of strengthening is realized in minerals, in the structures of which ions of metals (including lead) occupy neighboring structural positions sharing common ligands.

It is fully evident that the role of lead ions sharing common ligands with ions of transitional metals also resolves to the heightening of the degree of hybridization of the p orbitals of oxygen and d orbitals of the ion of the transitional metal in the Pb–O–Me complex, a consequence of which is a partial removal of the inhibition on the parity of the d–d transitions of the Me^{n+} ion and a strengthening of the intensity of the latter.

Thus, lead can be examined in the capacity of a catalyst of color, on the whole determined by ions of transitional metals. It is probable that ions of iron cannot be considered an exception, and in the spectra of the Pb–O–Fe complex, we are justified in expecting an increase in the intensity of the bands of d–d transitions of Fe^{3+} or Fe^{2+} ions. Precisely from these positions, the analysis of the reasons for amazonitic color and the basis of its model appears the most convincing.

First and foremost, it follows to state that serving as the precursor site of amazonitic color are the tetrahedrally coordinated Fe^{3+}_{IV} ions, with which is associated a wide absorption band in the 610–670 nm region in the optical spectra of feldspars. In this sense, amazonite can be compared with amethyst; the precursor sites of color of the latter, while considered Fe^{3+} ions, are contained in the mineral in highly insignificant quantities (as low as 0.n%). However, even the relatively high (~1%) content of the Fe^{3+} structural impurity in feldspars, as for example in ferro-orthoclase, still cannot lead to the appearance of amazonitic coloration: in the best case, a pale-green tone can be observed in certain microclines.

With the isomorphic entry of lead ions into the structure of iron-containing feldspar, it follows to expect an increase of the intensity of the 610–670 nm absorption band along with a corresponding strengthening of bluish-green tones of the color. However, such an effect is possible only in the case of the formation of exchange-coupled pairs of Fe^{3+}–O–Pb^{2+}, i.e., with the entry of lead atoms into the strictly defined structural positions sharing common ligands with tetrahedral positions occupied by Fe^{3+} ions. It is fully evident that this situation is realized predominantly in the ordered structures when Fe^{3+} ions, isomorphically replacing Al^{3+} ions, are situated in the defined tetrahedral positions, in part in the tetrahedra T_{1o} of the microcline structure.

Ions of Fe^{3+} occurring in the structure of microcline are the consequence of a much higher (by comparison with the replaced Al ions) share of covalence of their bond with ligands, are characterized by a less effective charge than that of Al^{3+}, and attempt therefore to surround themselves with diamagnetic cations of higher valence. In their turn, Pb^{2+} ions possessing a superfluous positive charge likewise prefer to be situated in close proximity to the ions of transitional metals that are conducive to lowering the effective charge of the high-valence ion on account of the exchange interaction.

The latter circumstance signifies that the lead ions formally occupying the position of K^+ (Fig. 5.4) will have the shortest bond with those ions of oxygen in the coordinating polyhedra PbO_{10}, which surround tetrahedrally the coordinated ions, that is, if in common microcline $K-O_{A2}$ is such a bond, then in amazonite it will be $Pb-O_{Co}$ (Fig. 5.5).

Such a distribution of ions Pb^{2+} in microcline structure defines the general direction of the overlap of wave functions of electrons participating in the exchange of atoms, that is, the correlated $Pb-O-Fe^{3+}$ complex. As is seen from Fig. 5.4,

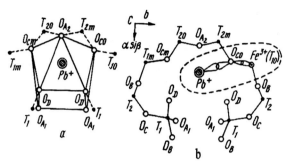

FIGURE 5.4
Coordinating polyhedron surrounding the Pb^+ ions (a) according to the data of A.S. Marfunin, and a model of a center of color (b) in amazonite: O_A, O_{A2}, O_B, O_C, O_{Co}, and the like (non-equivalent oxygen ions).

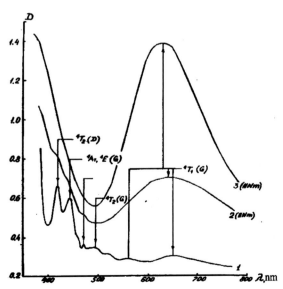

FIGURE 5.5
Optical spectra of absorption of (1–2) microcline (1: iron-containing, 2: green barium-containing); (3) blue amazonite; the arrows indicate the position of the absorption bands and provide their interpretation ($^4T_1$ (G), etc.).

the vector of the dipole moment of exchange-atoms is oriented predominantly along the crystallographic axis; accordingly, the projection of the bond Pb–O–Fe on this axis, and equally on the axis of optical indicatrice Ng, will be greater than is explained by the high intensity of the 635–650 nm amazonitic band in Nm-polarization. It is fully evident that the exchange interaction Pb–O–Fe^{3+} will be particularly effective if lead (according to Ref. [30]) forms in the structure of amazonite single-charge Pb$^+$ ions. In this case, the Pb$^+$ ion has an electron configuration of 6s^2 6p^1, that is, it possesses one electron on the valence p orbitals, which promotes the ancillary delocalization of electron density around the ions participating in the exchange, which are accompanied by an effective electron transfer from one of the cations to another by means of the ligand orbitals. In summary, the covalence of the bond in the Pb$^+$–O–Fe^3 complex fundamentally increases, which promotes an equally sharp increase in the intensity of the transition $^6A_1 \rightarrow {}^4T_1$ in Fe^{3+} ions, that is, the manifestation of band II.

On a similar basis rests the confirmation that the exchange correlation of lead and iron can also turn out to be fairly strong with the presence of a hydroxyl group in the capacity of a substituent for oxygen (Pb–OH–Fe^{3+}). In any case, the experimental data offer fairly convincing proof for the direct dependence of the intensity of amazonitic color upon the content of structural water [28].

At this point, it is appropriate to consider the term amazonitic band, which is applied to the spectra of absorption particularly of amazonites, that is, of potassium feldspars possessing a fairly intensive bluish-green color. It is fully possible that the pale-green color of other feldspars, including green orthoclase and other amazonite-like varieties of potassium feldspars as well as plagioclases, can be connected with other centers of color, for example, with those same exchange-coupled Me–O–Fe^{3+} pairs, in which in the capacity of a catalyzing ion participate other atoms, in part Fe^{2+} (especially in plagioclases) or Ba^{2+}. In this connection, it is appropriate to recall the observation of F. Frondel regarding the green color of the barium varieties of orthoclase. The spectrum of absorption (Nm-polarization) of pale-green barium-containing (1.76% BaO) microcline from the camptonites of the Azov region described in 1974 by A.A. Val'ter and G.K. Eremenko (see Fig. 5.5) is practically analogous to the spectrum of pale-colored amazonite, that is, it is characterized by a heightened intensity of the band of transition $^6A_1 \rightarrow {}^4T_1$ in Fe^{3+} ions. This permits the conclusion that the multi-charge ions of heavy elements (lead and barium) that enter into the structure of Fe^{3+}-containing feldspars render a fundamental influence on the spectroscopic parameters of the tetrahedral Fe^{3+} complexes. However, the manifestation of particularly amazonitic color is caused by only the formation of single-charge Pb$^+$ ions (electron centers of ions with anomalous valence, according to the classification of A.S. Marfunin), probably in combination with a hydroxyl. Accordingly, the exchange-coupled complex Pb$^+$–(O, OH)$^-$–Fe^{3+} can be classified as an amazonitic center of color. The authors et al. [39,46,47] have shown that the EPR signal behavior of the Pb$^+$ electron center after heating and irradiation can be well correlated with the behavior of the amazonite absorption bands at

625 nm. It is probable that in the alternative model could also participate this electron center.

It is understood that the saturation of color of an amazonite is determined by the concentration of the $Pb^+-(O, OH)^--Fe^{3+}$ pairs, that is, even with a singular structural state (degree of order) of the matrix, it depends as a minimum upon three variables: the content of relatively unstable electron centers of Pb^+, the concentration of Fe^{3+} ions in the T_{1o}-positions of the structure of potassium feldspar, and the content of the OH-group. It is not improbable that the formation of exchange-coupled $Pb^+-(O, OH)^--Fe^{3+}$ pairs by a known measure control the presence of Rb^+ and Tl^+ ions. Therefore, even in the presence of the alternative model examined above of the center of color of amazonites, in each concrete case marked correlations between the color and structural-chemical particularities of the studied samples appear highly problematic precisely in consequence of the multi-causality of the phenomenon (amazonitization) on the whole, and in amazonitic color in part.

In principle it is not improbable that a certain contribution to the 625–650 nm absorption band in the optical spectra of amazonites can introduce $Fe^{2+}_K-Fe^{3+}_{Al}$ pairs, supposing the possibility of the entry of Fe^{2+} ions in the position of potassium in the microcline structure. However (according to Ref. [28]), Fe^{2+} ions preferentially replace Ca^{2+} ions in plagioclases and the Fe^{2+} content in the latter increases with an increase in the content of anorthite molecules. These researchers did not report any notable heightening of the intensity of the absorption band in the region of 16,000–17,000 cm^{-1} (~600 nm) in the spectra of feldspars containing simultaneously variously charged iron ions. These results and the observations of the author et al. [47] regarding the particularities of the optical absorption of feldspars do not allow the proposal that the exchange-coupled $Fe^{2+}-Fe^{3+}$ pairs play a fundamental role in the formation of the amazonitic absorption bands.

In concluding the discussion of the nature of the 625–650 nm band, it is necessary to note once more that in many cases the so-called amazonitic band is a superposition of two absorption bands of differing natures, as follows from experiments on the thermal decoloration of the studied samples. Moreover, a fundamental influence is rendered on the tonality of the color of amazonite by a complex of short-wave absorption bands, as well as those contingent upon various reasons necessary to evaluate at least on the present level of knowledge on the nature of absorption centers in feldspars.

Band III (720–740 nm) represents one of the absorption bands of the hole centers of color. Its intensity is correlated with the intensity of band I; moreover, it is annealed within the same temperature interval (200–300 °C) as are the majority of the UV bands, at the least simultaneously with the 360–380 nm band.

It is fully probable that this absorption band is analogous in terms of its nature to the bands with E 13,500 cm^{-1} (~740 nm) in the spectra of certain natural sanidines and orthoclases, where (according to Ref. [48]) it is connected with centers of color of type O^-. Its high intensity in the optical spectra of amazonites can be contingent upon the elevated concentrations of similar centers forming as the result of the heterovalence isomorphism $K^+ \rightarrow Pb^{2+}$ and the subsequent formation of Pb^+ electron capture centers. Far from incidentally in the spectra of X-ray luminescence of amazonites is observed a clear direct correlation of the bands of emission of the centers of Pb^{2+} (~280 nm) and O^- (~480 nm; see Fig. 4.17).

It appears, likewise, indisputable that band III occurs in amazonites with a fundamental predominance of lead over iron; with values of $Pb/Fe > 15–20$, it is consistent in terms of intensity with the band of the exchange-coupled pairs $Pb^+–(O, OH)^-–Fe^{3+}$ (band II). As likely as not, the impurity lead ions not participating in the exchange-coupled pairs in the form of Pb^+ ions enter the local environment of the centers of O^-, promoting the polarization of s and r orbitals of oxygen ions interacting with lead ions, in which there is a redistribution of electron density and a corresponding increasing of the intensity of the optical transitions $2s \rightarrow 2p$ in the hole centers of type O^-. We add that precisely in amazonites have been observed the maximum concentration of centers $Pb–O^-$ or $Pb–O^-…x$ [20].

It follows to comment on the observations stated in [28] on the participation of Pb^{3+} ions in the coloration of amazonites.

With all possible in the structure of microcline modes of charge compensation with the heterovalent isomorphism, the occurrence of high-charge lead ions appears least probable; it is natural to suppose that with the isomorphism $K^+ \rightarrow Pb^{2+}$, a lowering of entropy of the system is carried out by means of the transition of lead ions into the single-valent state and, as has already been noted, their implication in the electron exchange with ions of transitional metals, in part of iron.

We note as well that the Pb^{3+} ions identified by A.N. Platonov and others [19,20] through EPR research data in certain natural samples of pink calcite are compensated by the formation of corresponding electron centers of CO^{-3}; precisely these centers, and not Pb^{3+} ions, cause the 500 nm absorption band in the optical spectra of such samples and, correspondingly, the pink color of the latter, fully disappearing with heating of calcite to 150 °C.

Band I (~360–380 nm) is not always individualized in the optical spectra of amazonites; it is noted only in relatively transparent samples not subjected to subsequent changes. For the overwhelming majority of amazonites in the optical spectra within the interval of wavelengths <400 nm has been reported a solid, short-wave absorption connected, in many cases, with the presence of a dispersed phase of the impurity of oxides and iron hydroxides, which are the products of the disintegration of the solid solution $KAlSi_3O_8–KFeSi_3O_8$.

In pure form, an analogous ~360 nm (27,500 cm^{-1}) band has been observed in the spectra of natural sanidines, and those artificially colored by X-ray irradiation, having a smoky color. Detailed research of the orientation and temperature dependence of these bands, as well as of the photoconductivity, EPR, and luminescence of smoky sanidines, has permitted the linking of the intensive 360 nm absorption band with the hole center represented by the hole localized in two equivalent oxygen ions in the cluster of three aluminum ions (the tetrahedra $T_1-T_2-T_1$).

The orientation and temperature dependence of band I in the studied optical spectra of amazonites permit the proposal that a certain contribution has introduced into its appearance one of the transitions in the hole centers of type O^- (Al–O^-–Al).

It, likewise, cannot be left unmentioned that there is an intensive ~380 nm absorption band noted in the optical spectra of certain transparent jewelry varieties of iron-containing feldspars: orthoclase, sanidine, oligoclase, andesine, and labradorite. This band is described by the transition $^6A_1(S) \rightarrow {}^4E(D)$ in the tetrahedrally coordinated Fe^{3+} ions and is accompanied by less intensive 420, 445, and 550 nm bands that are associated with other transitions in Fe^{3+} ions, in the aggregate causing the yellow color of feldspars. As has already been noted, certain of these bands occur in the optical spectra of amazonites only after the heating of the latter to 400–500 °C, that is, after the thermal destruction of the exchange-coupled pairs $Fe^{3+}-Pb^+$.

In this way, band I is likewise non-elemental in its formation—participate finely dispersed oxides and hydroxides of iron, as well as hole centers of type O^- (Al–O^-–Al) and $Fe^{3+}O_4$ centers.

In connection with the proposed models of centers of color in amazonites, we will examine, in conclusion, the processes of thermal decoloration and partial radiation-induced restoration of color.

One of the components of the exchange-coupled chromophore complex is the electron center Pb^+, thermally stable up to approximately 500 °C. Its ionization within the temperature interval 400–500 °C leads to the transition of the lead ion to the divalent state and correspondingly to the change of the character of the bonds in the $Pb^{2+}-(O, OH)^--Fe^{3+}$ complex, that is, to a decrease in the degree of hybridization of the orbitals of oxygen and iron atoms and, in consequence, to a sharp decrease in the intensity of the amazonitic absorption band. It is fully possible that at these temperatures a certain content of Pb^{2+} ions is displaced from the initial positions and is diffused toward other defect elements of the structure (cationic vacancies, for example), causing the nonreciprocal destruction of a defined part of the exchange-coupled complexes.

With subsequent artificial radiation treatment of amazonite, in its structure arise various centers of color, above all of type O^-, which conditions absorption throughout practically the entire visible range of the spectrum (see Figure 4.15). In part, in the spectra of irradiated specimens (particularly blue amazonites) is fairly distinctly expressed the band in the region of 540 nm (18,000 cm^{-1}), detected in the spectra of smoky sanidines with natural color or color strengthened by X-ray irradiation. Simultaneously, absorption increases sharply in samples within the

700–800 nm range; evidently, the Pb^{2+} ions remaining in their former positions capture electrons with the formation of Pb^+ centers with a restoration of the electron state of the exchange-coupled complexes that predetermine the appearance of the amazonitic absorption band.

Furthermore, it is interesting to note that the process of the formation of exchange-coupled pairs, and more so one of the components of a pair (the Pb^+ ion) in different samples can be carried out in different ways. Thus, in green amazonites with high content of lead impurity and a clearly expressed 720–740 nm absorption band after irradiation arise hole centers of type O^-, including $Pb–O^-$ centers; after heating the irradiated samples to 200 °C, the 720 nm absorption band dominates in their spectra (see Fig. 4.15). With extended heating to 300 °C, $Pb–O^-$ centers are destroyed with the formation of Pb^+ electron capture centers entering the exchange correlation with Fe^{3+} ions, which occurs with the increase of intensity of the absorption band of the $Pb^+–Fe^{3+}$ pairs in the 600–650 nm region. In blue amazonites, where the content of lead and iron impurities are fully commensurate, the formation of Pb^+ centers and correspondingly of $Pb^+–Fe^{3+}$ pairs proceeds immediately with X- or γ-ray irradiation.

The examined model of the center of color satisfactorily explains the loss by amazonite of the capability to restore color after its high-temperature (>800 °C) annealing. Evidently at these temperatures occurs a nonreciprocal diffusion of Pb^{2+} ions from the initial positions and a full destruction of the exchange-coupled complex.

All knowledge accumulated at present on the properties and particularities of amazonite can be summarized by the following points. In the color of amazonites participate as a minimum two centers of color responsible for the 625–640 nm (16,000–15,600 cm^{-1}) and 720 nm (~13,900 cm^{-1}) absorption bands, which are annealed at respectively 450–500 and 300 °C. The relative concentration of these centers determines the tonality of color of amazonite— from practically purely blue (the sharply dominating 625 nm band) up to the dense, grass-green and even yellowish-green (the significant impurity of the 720 nm band).

The indicated absorption bands and, correspondingly, the color of decolorized samples are restored with subsequent radiation treatment (X- or γ-ray irradiation). With a deeper heating of the samples (>600 °C), amazonitic color is not restored by irradiation.

Serving as a constant chemical characteristic of amazonite is the presence in them of lead impurity, the content of which varies from (0.01 to 2–3%). Moreover, the increase of the concentration of lead causes an increase in the intensity of the 720 nm absorption band and a corresponding increase of the role of green tones in the coloration of amazonite.

Blue tones in the color of amazonite predominate in samples with high structural order, that is, in maximum microclines ($\Delta_r > 0.98$) not contained in the same notable sodium impurity. The green tones of color are characteristic for samples with partially disordered structure (with the impurity of orthoclase domains). The 720 nm absorption band in pure form (without the 625 nm band) determines the green color of orthoclase from Broken Hill, Australia, containing 2% lead impurity [46,47].

The color of amazonite (more precisely, its blue component) remains practically unchanged with heating of the samples in a water medium in an autoclave up to 800 °C. This circumstance has become an incentive for further research and conclusions on the role of structural water in the coloration of amazonites [47].

The facts enumerated above were necessary to take into account on the basis of a more or less universal model of centers of color of amazonites. In this connection, the question is justly posited of a delineation of the concepts of amazonite and amazonite-type or amazonite-like feldspars [47,50].

According to the definition of the author et al. [47], which we fully share, as particularly amazonite should be considered the high-ordered lead-containing greenish-blue microcline, the color of which is determined by the 625–640 nm absorption band, signified henceforth as the amazonitic band (AB). That is, regarding the origin of the color of amazonite, it follows first and foremost to propose the nature of centers of color responsible for the appearance in the optical spectrum of amazonite of the mentioned absorption band. It follows to designate as the orthoclase band (OB) the auxiliary 720 nm absorption band that is contingent upon the impurity of monoclinic domains and determines the color of lead-containing orthoclases. Correspondingly, feldspars with radiation-induced green color and an optical spectrum dominated by the 720 nm absorption band are classified as the amazonite-type or amazonite-like (according to the analogy, for example, with amethyst and amethyst-type quartz, the centers of color of which are distinguished from those in true amethyst).

Regarding the absorption centers causing the 625–640 nm band (AB), there are various observations. Already in 1976 the author et al. [12] connected the color of amazonite with the paramagnetic center Pb^+, and in the research [28] was abandoned the preference toward a Pb^{3+} center (iso-electron with a Tl^{2+} ion) and proposed the active participation of structural molecules of water in the capacity of catalysts of valence transformation of lead ions. On the basis of theoretical calculations, A. Julg [29] proposes that the 625 nm (~2.0 eV) absorption band is connected with Pb^{3+} ions, while the 720 nm (~1.7 eV) band is connected with Pb^+ ions. In result of careful analysis of the EPR-spectra of amazonites, Petrov et al. [48] arrived at the conclusion of a connection of the AB with the exchange-coupled pair $[Pb^+-Pb^{2+}]^{3+}$, represented by variously charged lead ions in neighboring M-positions of the structure. In the opinion of the latter, the formation of such centers is possible only in high-ordered (maximum) microcline. The author et al. [47] with the research of the spectra of emission and the excitation of lead ions in amazonites has proven the presence in the latter of variously charged Pb^{2+} and Pb^+ ions; accordingly, they have put forth a model of a center of color in amazonites that proposes an exchange interaction of Pb^{2+} (Pb^+) and Fe^{3+}_T ions. In Table 5.1 are summarized the data on the possible mechanisms

Table 5.1 Origins of Color of Amazonites (The 625–650 nm Amazonitic Absorption Band, AB)			
Process	**[28]**	**[48]**	**Alternative model**
Isomorphism	$Pb^{2+}_{MA} + H_2O_{MB} \rightarrow 2K^+$	$Pb^{2+}_{MA} + Pb^+_{MB} \rightarrow 2K^+$; $Al^{3+} \rightarrow Si^{4+}$ in the immediate environment of the M_A-position	$Pb^{2+} + V_k^- \rightarrow 2K^+$
Ionizing radiation	Radiolysis of H_2O with the formation of H^0 in the M-position and of $OH^0 \rightarrow O^{2-}$ in the T-position	Radiational stimulation of the diffusion of Pb ions	Formation of free electrons e^- and of holes e^+
Formation of centers of color	Reduction of Pb^{2+} ions by atomic H^0: $Pb^{2+} + H^0 \rightarrow Pb^+ + H^+$	Ordering of multivalent Pb ions with the formation of exchange-coupled $[Pb^{2+}_{MA} - Pb^+_{MB}]^{3+}$ pairs	Capture by Pb^{2+} ions of an electron or hole with the formation of two types of color centers: 1. $Pb^+ - (O, OH)^- - Fe^{3+} -$ exchange-coupled complex or 2. Pb^+ Electron center
Decoloration (heating up to 500 °C)	Diffusion of protons H^+: $Pb^+ + H^+ \rightarrow Pb^{2+} + H^0$	Thermal diffusion of Pb ions	Thermal destruction of the electron center Pb^+ or exchange-coupled complex with the formation of Pb^{2+} ions

of lead isomorphism in the structure of microclines, as well as the formation and transformation of centers of color with radiation treatment.

As we see, the cited models of centers of color in amazonites are fairly contradictory; each of them fails to take into account all of the above-noted particularities and properties of amazonites. It is not improbable that amazonites of various granitoid formations are characterized by various mechanisms of coloration dependent upon the geochemical particularities of amazonite-containing microclines (Table 5.1), and that in principle, a universal model of centers of color in amazonites does not exist. It is far from incidental that among amazonites (of the blue varieties) in terms of intensity, width, and spectral position of the AB have been identified types B and T [28], model deposits of which—Lake George (Colorado, USA) and western Keivy (Kola Peninsula, Russia)—are associated with variously aged granitoid formations, Hercynian and Proterozoic, respectively.

As has already been noted, the 720 nm absorption band (OB) that causes the green color of lead-containing orthoclase from Broken Hill (Australia) and the green hues of color of amazonites, and fully anneals at 300 °C, has a nature different from that of the centers of color in amazonites. In terms of its own properties, this band is close to the bands that determine the green radiation-induced color of certain microclines ($15,400–16,100\,cm^{-1}$), but are shifted to the long-wave region of the spectrum. Most probably, the 720 nm band is connected with the hole centers of type $AlO^{4-}{}_4$, associated with Pb^{2+} ions in the M-position of the structure. We note that a similar band has been observed in the optical spectra of certain green plagioclases [29] where Al-centers can be associated with Cu^{2+} ions.

Ostrooumov et al. [47] reported an increasing content of radioactive elements (U, Th, Rb) in amazonite in relation to the paragenesis of amazonite with U and Th minerals. The electron–hole centers were formed in the amazonites in all genetic types wherein high concentrations of radioactive elements were found. The X-rays associated with radioactive decay could have caused formation of structural electron or hole defects, which later transformed in the centers of color in the crystals of the amazonite. Amazonites are always characterized by the presence of radioactive elements. This fact confirms the supposed participation of the radioactive elements in the formation of the centers of color in the crystals of the amazonites. We must stress the fact that as early varieties of the microclines were found in other granitic and pegmatitic deposits along with the amazonite, we could observe only the hole Al and the structural Fe centers, which produced the appearance of the complex band with its maximum in the near-UV region.

As stated above, the different colors of amazonite are caused mainly by Pb electron and hole centers that are produced by the reaction of radiation-induced Pb^{2+} with the radiation-induced products of dissociation of the structurally bound H_2O [47].

The change of color of minerals caused by irradiation is, in most cases, due to the formation of one or more color centers. In some minerals, this phenomenon may be accompanied by a change of valence of an element, which is in accord with the following reaction [46]: Amazonite $(Pb^{2+}) \rightarrow$ irradiation with X-ray or γ-rays $\rightarrow Pb^+$ electron center or O–Pb hole center. In addition, structurally bound H_2O, stimulated by irradiation, as shown in the FTIR-transmission investigations, may also take part in this reaction. Ionizing radiation, either X-rays or γ-radiation, dissociates H_2O molecules, forming H°- and OH°-radicals as the primary products of radiolysis. At the beginning of the amazonite-formation process, the OH° radically transformed the Pb^{2+} present in amazonite to Pb^+. The H_2O is further regenerated through the reaction of the OH° radical with a hole center (O–Pb) and the subsequent recombination of OH^+ and OH^- [47]. It is very probable that $^{IV}Fe^{3+}$ also participates in this reaction as well as in the formation of exchange linked Pb^+–O–Fe^{3+} complexes.

Thus, summarizing all of the above, it can be proposed that there exist the following models of centers of color of amazonite (in order of their decreasing significance):

1. Pb^+–(O, OH)$^-$–Fe^{3+}: exchange-coupled complex or/and Pb^+ electron center (band II–AB)
2. O^-–Pb^+: hole center in oxygen, connected with electron center in the lead ion (band III–OB)

3. Fe_2O_3 ($Fe_2O_3 \cdot nH_2O$): finely dispersed oxides (hydroxides) of iron (band I)
4. $Al-O^--Al$: hole center in oxygen, associated with an aluminum impurity (band I)
5. Fe^{3+}_4 or $Fe^{3+}O_4$: tetrahedrally coordinated with the oxygen ion of trivalent iron (band I)

5.2 GENESIS OF AMAZONITE

A.E. Fersman was one of the first to express a range of original judgments on the genesis of amazonite [23]. Above all, he drew attention to the specifics of the geological conditions of the discovery of this mineral in the then-known deposits, pointing to a "highly frequent occurrence of amazonite in silicificated pegmatites associated with nepheline syenites, if not directly genetically, then in any event spatially; however, there doubtless occur deposits of amazonite absolutely independent from alkali magmas" [23, p. 32]. In his proposed classification of pegmatites, which takes into account all then-known finds of amazonite, the latter was noted in connection with the following types of pegmatites: (1) common and with cerium; (2) with REE; (3) boron-fluorine (muscovitic); and (4) fluorine-beryllium. In the opinion of A.E. Fersman, amazonite freely crystallizes from melt in the relatively late geophases of the pegmatite process, but the beginning of its formation belongs to the earlier and high-temperature geophase—the rise of graphic granite.

Somewhat later, new understandings of the origin of amazonite were developed by A.N. Zavaritsky: "Amazonite was formed by way of a change and a transformation into amazonite of microcline not previously possessing a green color" [6]. In his observations, this process, which he termed amazonitization, occurred in pegmatites after albitization and silicification (see Chapter 1).

The two indicated points of view on the origin of amazonite in essence have not changed, even after the discovery of new genetic types of amazonite-containing rocks: granites, hydrothermal veins, ongonites, and others. Additionally, there have appeared new interpretations of the concept of amazonitization as a synonym of late microclinization, that is, of a metasomatic deposit of microcline-amazonite [2,47].

Apparently, the first notions of the acid–alkaline conditions of the formation of amazonite in granites were put forth in 1957 by N.L. Plamenevskaia, who considered the presence of amazonitic granites as evidence of the acidic character of the ore-forming solutions during the period of amazonitization, which in her opinion comprised the culminating stage in the post-magmatic processes.

A contrary conclusion on the character of solutions that cause the secondary amazonitic coloration of microcline (their alkalinity) was formed in 1967 by L.G. Fel'dman with collaborators [2]. In their proposed sequence of the post-magmatic processes, amazonitization occurring at temperatures of 250–300 °C was also allocated a concluding role following after albitization, greisenization, and their accompanying ore mineralization. Moreover, these authors noted the possibility of a direct crystallization of amazonite from the solutions in the process of formation of hydrothermal quartz-amazonite veins.

The examined principal hypotheses on the origin of amazonite provide evidence of the highly vague and contradictory understandings of the given question. In this connection, the author et al. [47] have considered it necessary to harmonize the existing opinions and to express their own views on the particularities of the origins of this mineral, having noted that for the judgment of genetic aspects of the amazonite problem are required not only geological-mineralogical observations (without which the genetic understandings generally lose all sense), but also the comprehensive results on the experimental study of the crystal chemistry and properties of amazonite.

The author et al. [47] propose to examine the genesis of amazonite in the following sequence:

1. The general geological conditions of the discovery of this mineral, which will allow judgment of several particularities of its formation (formational types and subtypes of granitoids, types of amazonite-containing paragenesis, spatial–temporal correlation with other magmatic and post-magmatic processes)
2. The modes of formation of potassium feldspar with specific crystal chemical features—the set of impurities that create precursor sites and centers of amazonitic color, and the defined structural state favorable for its appearance
3. The genesis of amazonite color itself—the time, modes, and conditions of its development

Only after examining these questions can the particularities of the formation of amazonite be considered and further on this basis can judgment be expressed on the prospecting-evaluating significance of amazonitic feldspar.

We characterize the main particularities of the geological setting and evolution of amazonite:

1. There exist two maximums of manifestation of amazonite in nature: the subalkaline-leucogranite maximum and the alkaline-granite-pegmatite maximum. In the first case is noted a fully distinct growth of the degree of the development of the amazonitization process from the early phases of granites to the late phases with a maximum in pegmatites (stockscheiders) localized in peripheral endocontact parts of a plutons. By no means do all plutons of this formational type turn out to contain amazonite—only a part of those belonging to the fundamentally alkaline varieties and predominantly the relatively young and low depth. Among pegmatites of the alkaline-granite formation (western Keivy), the maximum of amazonitization can be observed at a certain optimal distance of vein bodies from the plutons of source granites, and its attenuation occurs as much on the side of an elevation of alkalinity of the environment (in the direction toward

the plutons of granites) as on the opposite side of strengthening of the role of acidic paragenesis. Furthermore, within the limits of the separate vein bodies, amazonitization accrues by measure of progression toward their central and lower zones. A strengthening is noted of its degree with an increase in the age of the pegmatites of this type.

2. Amazonitization is an evolutionary process consisting of three stages. The first of these—the pre-amazonitic (preparatory)—serves as a necessary stage for the development of the entire subsequent process. The next—the particularly amazonitic stage of blue coloration—gives way later to the stage of green amazonitic coloration. All three stages can either overlap or occur together within the limited extent of the separate vein bodies (western Keivy), or be divided in space, as for example in granites and their other genetic types. Most frequently, however, the process of amazonite formation[1] is represented by a single one of the enumerated stages: blue in the Il'menskie Mountains and green in the Broken Hill deposit. In the most fully manifested form, the particularly amazonitic stage is noted in one of the veins of the western Keivy pegmatite field.

3. Characteristic for each of the identified stages of amazonite formation are fully defined mineral paragenesis. In pegmatites of the alkaline-granite formation, in the first stage, besides the main rock-forming minerals (quartz, acid plagioclases, biotite, and microcline-perthites of common colors) are formed minerals essentially not containing a large quantity of volatile components or generally without them, but instead with high concentrations of titanium, iron, REE of the cerium group (ilmenite, titanite, magnetite, zircon, tscheffkinite, abukumalite, and others). In the second (blue) stage is noted the appearance of new (but, as previously, dehydrate) accessory minerals (gadolinite, fergusonite, columbite, and cassiterite), completed subsequently by the accumulation of a range of minerals with volatile components, mainly with the hydroxyl group (biotite, zinnwaldite, topaz, tourmaline, and others). In the concluding third amazonitic (green) stage and after it here alongside hydroxyl-containing minerals, a notable role begins to be played by fluorine-, lead-, yttrium-, and sulphur-containing accessory minerals (zinnwaldite, topaz, fluorite, genthelvite, plumbum-microlite, galena, and others).

In granites of the subalkaline-leucogranite formation, with the transition from the beginning to the concluding stage of amazonite formation is observed still a more distinct evolution of the composition of micas (from iron-magnesium to fundamentally lithium, from biotite to zinnwaldite and lepidolite), a significant accumulation of fluorine-containing minerals (initially topaz, and subsequently also fluorite), and the increase in concentration of accessory minerals with volatile components (from columbite-tantalite to pyrochlore-microlite).

---

[1]Amazonite formation: a general concept for the designation of various modes of formation of amazonite, including amazonitization (secondary coloration).

4. To each of the above-indicated stages of amazonite formation corresponds its own stage of perthite evolution: the pre-amazonitic stage corresponds to the initial segregation, the first amazonitic stage to the most complete segregation, and the second amazonitic stage to de-perthitization.

5. The analysis of the correlation of the chemical elements in potassium feldspars and amazonites of various stages of coloration provides evidence of the undoubted loss of some elements (sodium), the supply of others (potassium, rubidium, cesium, and lead), and the indistinct tendency (closer to loss) for the third (iron and silicon). The calculation of the balance of the loss and supply elements in each of these stages allows us to conclude that the pre-amazonitic stage should be characterized as the formation of potassium feldspar with vacant cationic (cation-deficient) sub-lattice on the account of the predominance of the loss sodium over the supply potassium and other alkaline cations. The maintenance of the electron neutrality of the lattice is possible by means of charge compensation (in all probability, at the expense of oxygen, by way of the creation of its hole centers or replacement by hydroxyl or fluorine).

6. Following from the paragenesis and particularities of the chemistry of amazonites can be noted the following sequence of the regime of acidity–alkalinity in the context of amazonite formation: the early moderately acidic stage (pre-amazonitic), the early alkaline (blue amazonitic), the acidic (inter-amazonitic), the late alkaline (green amazonitic), and the late acidic (post-amazonitic). In other words, in the complex bodies of pegmatites and granites can be identified two stages of amazonite formation (and their closely associated albitization) and three acidic lixiviation (in part, greisenization). Hence, the possibility remains clear for a multistage formation of ore (rare-metal, REE, and others) minerals associated with various stages of acidity–alkalinity, including those identified with amazonite formation.

7. Known and undoubted are the following conditions of the formation of amazonite: a) magmatic (ongonites, particularly granites and vein granite-aplites) and b) post-magmatic (stockscheiders and central zones of pegmatites; possibly certain crystals in plutons of granites and bodies of metasomatites, late hydrothermalites).

   The upper temperature boundary of the formation of the given mineral (ongonite phenocrysts) consists of approximately 650 °C, and the lower (hydrothermalite) limit is 250 °C. Accordingly, the following mode of its formation can be identified: primary crystallization from melts (in ongonites, granites) and solutions (in cavities of pegmatites and hydrothermalites); recrystallization (in granites, stockscheiders, certain pegmatites); metasomatic development of blastocrysts in granites and metasomatites; and ion-metasomatic replacement (secondary coloration, according to A.N. Zavaritsky).

8. Amazonitic color can be primary (crystallization) or secondary (post-crystallization).

We move on to an exposition of the authors' understandings of the genesis of amazonite, having in mind in the first order two principal variants of its formation: pegmatite (secondary ion-metasomatic coloration) and ongonite-granite-hydrothermalite (primary crystallization). We begin with the first of these.

Precursors to amazonite in pegmatites are potassium feldspars from graphic, apographic, or blocky zones. A necessary prerequisite for their fullest manifestation of amazonitization should be considered a fairly complete recrystallization with a wide distribution of blocky K-feldspathic and quartz zones and the disappearance of graphic and apographic zones. Most frequently arising are potassium feldspars of common colors with perthites of various stages of segregation. Such colors are associated with the independent phase of oxides and hydroxides of iron formed with the disintegration of potassium-sodium (with structural iron) feldspar in a relatively oxidizing environment. These potassium feldspars during the early moderately acidic stage (to which belong blocky quartz and biotite of the central zones) are subjected to decationization (essentially a loss of sodium uncompensated in full measure by a supply of potassium and other alkalis). The process of supply–loss of alkalis could have occurred with the condition of active participation of volatile and especially of hydroxyl groups. The latter, replacing oxygen, enable the maintenance of electron neutrality of the lattice of the feldspar (the same is possible with the formation of hole centers in oxygen). Precisely here is manifested the existence of a pre-amazonitic stage of altering of potassium feldspar.

The next, particularly amazonitic, stage of coloration is clearly reported already, not only in terms of loss of sodium from microcline (with the simultaneous development of albite I), but also in terms of a distinct supply of potassium, rubidium, cesium, and lead. Moreover, the rate of accumulation of potassium is most frequently the least, while that of lead in many cases is the greatest; those of rare alkalis are intermediate. For the given stage are established the following schemas of the rate of accumulation of the indicated elements: $Pb > Cs > Rb > K$ (western Keivy); $Pb > Rb > Cs > K$ (Il'menskie Mountains, mine 70); and $Rb > Cs > Pb > K$ (the same, mine 77). These data are evidence of a gradual increase of alkalinity in the environment of the formation of blue amazonite. We note as well during this stage a strengthening of the magnitude of the hydroxyl groups, which transfer alkaline and other rare metals (beryllium, niobium, tantalum, and tin) and participate in the structure of the exchange-coupled complexes $Pb^+-(O, OH)^--Fe^{3+}$ (centers of blue color, see Chapter 4). These play a catalyzing role in the phenomenon of Si-Al-Fe order–disorder, which enables a convergence and consolidation in the complexes of lead and iron atoms. We suggest that this stage is closely genetically associated with the late, more intensively manifested segregation of perthitic ingrowths; therefore, the most intensive blue colors are observed in the coarse perthitic varieties of amazonitic potassium feldspar.

Corresponding with the proposed schema of the evolution of amazonite formation is the subsequent stage: the acidic (greisenic). During this stage, the previously formed amazonite either generally cannot be maintained as a potassium feldspar as a result of hydrolysis and lixiviation accompanied by the development of cavities, or it loses color to some degree (the phenomenon of de-amazonitization) on account of the partial destruction of the centers of blue color and potential loss (decationization) not only of sodium, but also of potassium and iron. Furthermore, the typical paragenesis is formed of accessory greisenic minerals: topaz, beryl, tourmaline, lithium-iron mica, and others.

With a further drop in temperature and a change of the regime acidity–alkalinity, a transition is also carried out to the following stage of amazonite formation (the late alkaline stage), characterized by the appearance of centers of green color (O^-–Pb^+). The brightest sign of such amazonites, besides green color, is a distinct de-perthitization of potassium feldspar, which provides evidence of a clear loss of sodium, which in turn enters the composition of albite II. The loss of sodium (and partially of iron) accompanies a yet stronger than that of the first stage of amazonite formation enrichment of microcline-amazonite—first of all in lead, and then in other elements according to the schemas: initially $Pb > Rb > K > Cs$ and subsequently $Pb > Cs > Rb > K$. This process is accompanied as well by a certain Si-Al disordering and simultaneous change of cationic order. Highly typical for the described stage are reductive conditions of the environment of mineral formation, reported in terms of the conversion of Pb^{2+} into Pb^+, the sharp increase of the Fe^{2+}/Fe^{3+} relationship, and the appearance of typical associations of sulfides and native elements. The character of an environment with high activity of rare alkalis, lead, and volatile elements (predominantly fluorine) is confirmed by the experiments of K.K. Matveev, V.N. Florovsky, and others on the extraction of hydrocarbons from amazonite, as well as by the data of L.N. Kogarko on the accumulation of hydrocarbons in the low-temperature gas phase.

The stability of the O^-–Pb^+ centers arising during this stage is explained not only by the presence of hydroxyl groups that play a fundamental role in the first stage of amazonite formation, but also by the appearance of fluorine (the structural position of which is not yet clear). It can be proposed that if even this element is not included in the composition of centers of green color, then at least it enables their appearance (the supply of necessary elements) and maintenance (the creation of a reductive regime). It follows to add that during this stage of amazonitic coloration are noted relatively elevated concentrations of radioactive elements (thorium, uranium, rubidium, and potassium) and, as a consequence, of the corresponding radiogenic isotopes (lead, strontium, and calcium). It is likely that radioactive decay can give rise to additional precursor site lead ions, as well as electron–hole centers, which can participate in the formation of centers of color.

The very last low-temperature stage, the acidic stage, occurs in the development of the following paragenesis of minerals: fluorite (yttrofluorite), aluminofluorides, sulfides (galena), the complex of fundamentally yttrium REE accessories (keiviite, vyuntspakhkite, and others), and native elements (bismuth and antimony). During this stage, fine cavities of lixiviation are formed on account of the destruction of albite or green amazonite, filled in by the latest paragenesis of minerals (finely dispersed clayey minerals with iron hydroxides), while the late gray quartz is represented by skeletal metacrystals (western Keivy).[2]

The cited schema of the genesis of amazonite in pegmatites of the alkaline-granite formation is generalized and proposes a predominant formation of amazonite by way of secondary coloration. This does not exclude, however, the possibility of another mode of formation of amazonite—by way of primary crystallization in the cavities that arise during some stage of acidic lixiviation. In such cavities correspondingly can be observed crystals with initial amazonitic color of various tones, from blue to dirty-green. Thus are possible widely differing instances of mutual distribution of colored and non-colored zones and sectors with amazonitic and non-amazonitic colors (see Section 4.1). It follows to accent attention on the fact that the evolution of amazonite formation is most frequently not expressed in full form during all the stages enumerated above, but it can begin or break off in any of them. In full form, evolution can be observed in pegmatites of western Keivy, whereas in the Il'menskie Mountains, amazonite formation is concluded during the first blue stage, and in the Broken Hill deposit, evidently, only the last (green) stage of amazonitic coloration receives distribution.

Transitioning to an examination of the genesis of amazonite in granites, it is necessary above all to note the principal congruence of the evolution of amazonite formation in pegmatites with that of granite plutons. In the latter, in the same sequence as in pegmatites, occur as a minimum two alkaline stages of amazonite formation, with a division of their stages of acidic lixiviation. In concrete geological situations can be developed, as well as in pegmatites, not all stages. Corresponding to these stage can be seen some general features of the paragenesis of minerals and the evolution of their chemistry. Thus, there are distinguished two generations of albite in association with variously colored amazonites; a distinct change is noted in the composition and properties of micas from the early biotites to the late lepidolites, as well as a strengthening of the role of fluorine in terms of the increase of the content of topaz, to which subsequently is added fluorite and micas rich in fluorine.

The visually observable diversity of amazonites is connected with the diverse character of the differentiation of granites and their other genetic types on the one hand, and of pegmatites on the other. Early portions of the granite melting can produce vein spalling in the form of ongonite dikes, the establishment of which occurs in the subsurface conditions and in which occasionally occur amazonitic phenocrysts. The first portions of particularly granite melting are formed already at depths of several kilometers. Granite phases here can contain

---

[2]A five-stage formation of amazonite (three acidic and two alkaline stages) does not correspond to the traditional conceptualizations of a three- or four-stage regime of acidity–alkalinity of the post-magmatic processes.

amazonite of bluish colors. Late albitite and possibly pegmatoid formations enclose green and bluish-green amazonites, and the latest green varieties of potassium feldspar are intrinsic principally to hydrothermalites situated in the supra-intrusive zones.

In pegmatites, as has been noted, late green varieties of amazonites are characteristic for the lower zones of the veins, and the early bluish varieties are characteristic for the upper and peripheral sections. More intensive colors of amazonites in pegmatites (with the condition of a fairly fully manifested amazonitization) are explained by a high concentration of centers of color in potassium feldspars of these rocks; whereas in granites and their other genetic types, fully fledged in a large extent, such concentration of amazonitic centers is probably not reached. Therefore, the intensity of color here is fundamentally lower. Precisely thus can be explained as well the different concentration of ore minerals, insofar as the development of amazonitic centers and deposition of ore minerals is associated with the same solutions.

Finally, we once more emphasize that in ongonites, a large part of particularly granitic rocks, granite-aplites, and hydrothermalites, the author et al. [47] allocate a leading role to the primary crystallization of amazonite, while not excluding entirely the role of recrystallization and metasomatism (for example, in giant-grained granite pegmatites and zones of metasomatites).

Concluding the discussion of genesis, we will concern ourselves with the question of the correlation of amazonite formation with other important stages of rock (albitization, greisenization) and ore formation. It is necessary to note the coexistence in the literature of highly contradictory evaluations of the spatial–temporal correlations of these phenomena. We will name the extremes of these conceptualizations: amazonitization as the earliest post-magmatic change of potassium feldspar preceding various stages of albitization and mineralization in pegmatites (according to A.Ia. Lunts), and amazonitization as the latest post-magmatic process of change of potassium feldspars in granites, superimposed on the preceding stages of albitization and greisenization and their accompanying ore mineralization (according to L.G. Fel'dman et al.).

Similar understandings of the place, time, and corresponding significance of amazonitization appear to be inadequately supported and to a significant degree simplified, failing to account for the stadiality of amazonite formation. From the authors' point of view, what is needed is a differentiated approach to amazonite in connection with its corresponding phenomena of albite formation, greisenic parageneses, and ore mineralization. Such an approach provides for the following: (1) the syngenetic nature and spatial connection of blue amazonite with early albite, and of green amazonite with late albite (keeping in mind that the maximums of occurrence of these processes are spread out in space); (2) the varying spatial situation of irregularly colored amazonites in the bodies of pegmatites and granites as well as within the limits of vein fields; (3) the mutually exclusive role of greisenization (and generally of acidic processes) and amazonite formation: the intensive development of acidic parageneses either destroys centers of color in previously existing amazonites, or destroys that amazonitic material entirely, extracting part of the volatile components in the creation of greisenic minerals; (4) the necessity of examining the interrelation of the evolution of ore-genesis with the proposed stadiality of amazonite formation; and (5) various spatial–temporal correlations of ore minerals and amazonites of early and late stages.

We will illustrate the above by way of several examples. In the alkaline-granite pegmatites of western Keivy, the most intensive greisenization (and more frequently silicification) is noted in the vein bodies of the near-endocontact, whereas the maximum of amazonitization occurs in pegmatites extending on the whole to the periphery of the field; a certain intermediary situation is occupied by pegmatite veins with a maximum occurrence of accessory ore minerals of somewhat different composition: beryllium, essentially tantalum, and yttrium. It is highly characteristic that early ore minerals (gadolinite) are more frequently associated with blue amazonites, while niobium-tantalum (manganocolumbite and plumbomicrolite), partially yttrium (xenotime, keiviite, caysichite, and others) and sulfide (galena) are associated with late green amazonites. However, the basic mass of REE mineralization is associated with the post-amazonitic acidic stage, where yttrofluorite, according to A.V. Voloshin et al. (1986), is an indicatory mineral. We emphasize once more that the maximums of amazonite and ore formation do not coincide spatially, for which reason typical amazonitic pegmatites should be considered as deposits of raw gemstones.

In amazonite-containing granites of Transbaikal and Mongolia, precisely as in pegmatites, the maximums of amazonite occurrences do not coincide spatially with the maximums of development of albitization and greisenization. In typical albitites of the upper parts of the plutons, amazonite (commonly bluish-green) is rare or practically non-occurring. Amazonite on the whole is likewise not particular to greisens of the supra-intrusive zone and to hydrothermalites; if amazonite does exist there, it most frequently has green hues of color. Characteristic for the fundamentally albitic granite rocks is the most intensive mineralization of a tantalum-niobium character (tantalite and microlite) with a maximum Ta/Nb ratio. With greisens is associated principally tin mineralization (cassiterite). In albite-amazonitic granites, where amazonite is distributed most widely (its color more frequently is blue or bluish-green), mineralization is represented primarily by intermediary members of the isomorphic solid solutions of pyrochlore-microlite and columbite-tantalite, with a Ta/Nb ratio and some amount of these rare minerals, somewhat less than that characteristic for albitites. The very earliest pale-blue and partially colored amazonites are commonly observed in relatively deep and early phases of amazonitic granites, with which is associated a poor, fundamentally niobium (columbite) mineralization.

Finally, we note the classic example of tin-bearing granites of the Ore Mountains, Cornwall, and France, in which amazonite either occurs extremely rarely (episodically) or is generally absent. On the other hand, no less bright representatives are known of granites of Kazakhstan with massive and intensively manifested amazonitic color, in which is noted a poor Ta/Nb mineralization, dispersed along the entire extent of the granite body. In such instances, amazonitic granite represents an industrial interest in the capacity of a handsome facing material for construction.

CHAPTER 6

Significance of Amazonite

6.1 TYPOMORPHISM AND PROSPECTING SIGNIFICANCE OF AMAZONITE

During the last few decades the typomorphism of minerals has been widely and effectively used in prognosis, exploration, and evaluation of mineral deposits. A special role in solving these problems is played by fairly widespread or rock-constituent minerals that can be formed in various conditions of formation in the earth's crust. The study of the typomorphic features of these minerals enables us to obtain unique information on genetic and prospecting-evaluation characteristics. Amazonite should be classified among such minerals. Its fairly significant distribution in various genetic and formational types of granitoid rocks, easy diagnostics, and formation during defined stages of mineralogenesis, along with its consequently distinct typomorphic features—these are all features that would recommend the given mineral as highly informative and useful in the detailed study of a range of geological questions and solutions of prospecting-exploratory problems [16,26,46,47]. It follows to emphasize that according to one of the known researches on pegmatites and granites throughout the world [26], in more than 80% of deposits, amazonite is associated with rare-metal and REE minerals.

During the last few decades, geologists and mineralogists have discussed the use of amazonite in the exploration for deposits of rare metals and REE, and there are different (both positive and negative) opinions regarding this problem. From our point of view, the correct response to all of this discussion is to be found in the detailed characterization of the geological setting of the amazonite, as well as in the proper investigation of its chemical composition, structural features, and properties with respect to each genetic type of granite. According to our data, this concrete geological setting influences the distinctive (or typomorphic) features of the crystal chemistry and properties of amazonite [9,16,47].

Amazonite: Mineralogy, Crystal Chemistry, and Typomorphism. http://dx.doi.org/10.1016/B978-0-12-803721-8.00006-8

As has already been emphasized in Chapter 3, in the main formational and genetic types of amazonite-containing rocks have been identified fully defined typomorphic paragenesis of rock-forming (occasionally secondary) and accessory minerals.

In granitoids of the subalkaline-leucogranite formation, such typomorphic mineral associations found among rock-forming minerals include quartz, microcline (including amazonite), albite, and lithium mica; among accessory minerals, they include topaz, fluorite, columbite-tantalite, pyrochlore-microlite, cassiterite, and monazite. In the later formations of genetic types of rocks of this formation (greisens, albitites, and quartz-amazonitic veins) is maintained the same paragenesis of rock-forming minerals; among their accessory minerals appear lead-containing plumbopyrochlore and galena.

In terms of concrete paragenesis of secondary and accessory minerals, the author et al. [47] have identified diverse specifics of the subtypes of the subalkaline-leucogranite formation. Characteristic for plutons of the greisen-albitite subtype is a significant scale of mineralization with a wide development of topaz, tantalite, microlite, Ta-cassiterite, and others. Practically oreless intrusions of the amazonite-granite subtype are distinguished by a somewhat different specific set of micas (protolithionite, zinnwaldite, muscovite), along with an insignificant distribution of columbite, cassiterite, fluorite, thorite, zircon, and tourmaline.

All of the indicated mineral associations in their most fully manifested form are reported predominantly in the peripheral and upper parts of granitic plutons; their discovery therefore indicates a low level of erosion of the intrusions.

Such a rock-forming association is particular to pegmatites of this formational type. Of accessory minerals in these rocks, alongside those noted for granites are observed minerals highly specific for them, such as polychromatic beryl and tourmaline (often tsilaisite), as well as manganese-apatite, and they are rich in manganese garnets (pyralspite).

In ongonitic rocks on the whole is confirmed the same leading typomorphic paragenesis; however, in these formations, lithium mica and topaz are developed to a lesser degree than in granites. The typomorphic paragenesis of accessory minerals is represented in ongonites by fluorite, garnet, zircon, columbite-tantalite, and cassiterite.

In metasomatites and pegmatites of the alaskite and alkaline-granite formations, lithium mica as a rule is absent. Most typical for other genetic types of the first of these formations are muscovite and biotite, while occurring more frequently in the alkaline-granite formation is biotite commonly associated with alkaline dark-colored minerals. In metasomatites, the latter are occasionally noted as well in galena, ganite, willemite, genthelvite, and cryolite. In terms of the composition of accessory minerals, metasomatites of the alaskite formation are distinguished somewhat from amazonite-containing rocks of other the formations. Characteristic for these are columbite, schorl, fluorite, and occasionally beryl. In the more ancient and deeper pegmatites of the given formation, the typomorphic association of accessory minerals is muscovite-beryllium with fluorite and uraninite, and in the young and less deep pegmatites, it is muscovite-beryl-topaz with thorite and fluorite.

In pegmatites of the alkaline-granite formation, the typomorphic paragenesis of accessory minerals changes quite fundamentally within the formation (from its early to late members): from predominating REE with subordinate development of rare-metal to the opposite relationship between them (see Chapter 2).

Typomorphic features of minerals-accessories of amazonite (in part, their colors and certain particularities of chemical composition) have already been discussed (see Chapter 3). In this section, we emphasize that in the rocks of all granitoid formations, a distinctive feature of such minerals is their enrichment by tantalum, provided that the concentration of this element, when reported in late generations of minerals, occurs in parallel with a strengthening of the degree of the amazonitization of potassium feldspars. Rare alkaline elements are accumulated analogously; the highest contents of them in each formation's development are established predominantly in the latest amazonite-containing paragenesis (principally in potassium feldspars and lithium micas). Frequently noted as well is the enrichment of fluorite by REE of the yttrium group (up to the appearance of yttrofluorite), of garnet and tourmaline by manganese, and of zircon by hafnium.

The enrichment and the analysis of numerous data on the geological setting of amazonite allow us to consider this mineral as a specific indicator of aureoles of a find of various genetic types of the above-indicated rare-metal-bearing granitoid formations. Among them the greatest distribution is characterized by the amazonite-containing rocks of the subalkaline-leucogranite and alkaline-granite formations; development of the first should be expected in the younger and less deep regions, and development of the second in the more ancient and deep regions. In other words, the discovery of amazonite in some paragenetic association potentially suggests the formational classification of concrete granitic plutons and pegmatite vein bodies. In the overwhelming majority of cases, the presence of amazonite in granites and pegmatites provides the basis for their classification within the subalkaline-leucogranitic and the alkaline-granite formations. Pegmatites and metasomatites of the alaskite formation are commonly characterized by a very weak local development of amazonite (more frequently with a low intensity of color) and are practically non-ore bearing. The obligatory consideration of the typomorphic mineral paragenesis of rock-forming and accessory minerals is conducive to the most precise definition of the classification of rocks to a particular granitoid formation type.

From the point of view of the author et al. [47], amazonite can be considered a mineralogical criterion in the indication of concrete formation subtypes of granitic plutons. Thus, in the amazonite-granite subtype of the subalkaline-leucogranite formation, potassium feldspar is represented principally by amazonite, whereas in the greisen-albite subtype commonly colored potassium feldspar as a rule predominates over amazonite.

It is important to emphasize once more that the plutons belonging to these subtypes are differentiated in terms of scale and degree of ore mineralization; the plutons of the greisen-albite type are considered more promising for rare-metal mineralization. Established for the latter, as has already been indicated, is a wider development of lithium micas (up to the appearance of mineral species with the maximum lithium content) and of tantalo-niobates with a high concentration of tantalum (tantalite and microlite). Present in the amazonite-granite subtype are micas of protolithionite-zinnwaldite various members (with lowered content of lithium); rare-metal minerals are represented only by columbite. The definition of the type and subtype of a granitic pluton, as well as the level of its erosion can be derived in terms of certain particularities of the crystal chemistry and properties of the amazonite.

A fairly high content (no less than 20–25%) of amazonite in granites serves as evidence of a low degree of erosion, insofar as amazonite according to geological observations is concentrated principally in the peripheral and upper zones of plutons. In a range of cases, pale-colored (blue) varieties of amazonites can be found also at the deeper horizons (up to 250–300 m). The presence of stockscheiders (pegmatoid veins) with amazonite also points to a rather low level of erosion of granite bodies.

With the discovery of amazonite in pegmatites, it is necessary to approach each concrete case with a differentiated evaluation of the level of erosion of the vein bodies. In other words, the stadiality of amazonite formation (see Section 5.2) should always be kept in mind along with the varying degree of occurrence of this process in pegmatite veins.

Finally, from the analysis of amazonite genesis (see Section 5.2) it follows that this mineral is a typomorphic mineral indicator of the alkaline (K, Rb, Cs) process of coloration of potassium feldspars in reductive conditions with relatively high concentration of Pb, Th, and U, along with volatile components (OH, F, B), within the temperature range 650–250 °C. The presence of amazonite in subsurface rocks serves as evidence of the potential identification in or at a certain distance from them of ore formations of a particular type, associated with one or several stages of amazonite evolution. Thus, distinctly differentiated in terms of the quantity of amazonite and the scale of ore mineralization are the inner and outer zones of granitic plutons (Transbaikal) or the pegmatite veins at various distances from alkaline granites (Kola Peninsula). Additionally, in a range of cases, amazonite occurs in close paragenesis with rare-metal and REE minerals (pegmatites of the Kola Peninsula and the Urals, metasomatites of Ukraine, and granites of Transbaikal and Mongolia).

We will consider the typomorphism of the crystal chemistry of amazonite. As follows from the results of Chapter 4, amazonite is a mineral concentrator of such rare elements as rubidium, cesium, and thallium. We especially emphasize the consistent presence in amazonites of concentrations higher than in other potassium feldspars of lead, thorium, and uranium, as well as of the volatiles OH, F, B, and others. These in essence are the main typochemical features of oxides and sulfide (above all, galena) mineralization. Observed in the younger and less deep pegmatites of this formation is a marked paragenetic association of blue amazonite ($\lambda = 491–494$ nm) with topaz, and of green and yellowish green amazonite ($\lambda = 540–570$ nm) with certain late tantalo-niobates (Urals).

In correspondence with the indicated changes of colors of amazonites has been established the fully defined evolution of the centers that generate those colors (see Section 4.4). To the initial stages of the process of amazonitization corresponds the formation of complex exchange-coupled complexes of lead and iron (the origination of "blue" centers of color and of absorption band II). By measure of the further development and strengthening of this post-magmatic process is observed a gradual increase of relatively low-temperature Pb^+ electron and oxygen hole color centers (the formation of "green" centers and absorption band III). In the end of the process on account of the increased concentration of hole centers and the strengthening of absorption band III, the green component of color begins to predominate. In other words, over the course of the process of amazonitization as the temperature and the concentration of structural iron decrease (with an increase of the Fe^{2+}/Fe^{3+} ratio), and as the content of lead in the mineral-forming environment in amazonite potassium feldspars increases, there occurs a regular redistribution of the roles of various color centers, expressed as a gradual decrease in the number of exchange-coupled complexes and an increase in the number of electron–hole centers.

A regular change in the same direction of the luminescence characteristics of amazonite on the whole confirms all of the above (see Table 4.16). Thus, as the process of amazonitization strengthens, there is an increase in the values of the ratio of the following centers of luminescence: O^-/Fe^{3+} (with a multidirectional change of the number of common centers) and Pb^{2+}/Fe^{3+} (mainly on account of the sharp increase of the number of Pb^{2+} centers). We draw attention to the fact that in terms of the concentration of these centers, there is a fairly sharp differentiation between the plutons of the two subtypes of the subalkaline-leucogranite formation: the maximum concentrations of the first are noted in the granite bodies of the greisen-albitite subtype (Transbaikal), while in the granites of the amazonite-granite subtype the second always predominate (Kazakhstan).

The identified dependences can be used also in the evaluation of the level of erosion of granite bodies, insofar as in amazonites from endocontact zones of intrusions that are paragenetic with rare-metal mineralization, the ratio of the concentrations of the indicated centers of luminescence reaches maximum values.

Finally, these dependences provide complementary information for the prediction of the sites of localization of ore mineralization, insofar as amazonites from rare-metal paragenesis are characterized by fully defined spectroscopic parameters: the highest ratio of Pb^{2+}/Fe^{3+} centers (up to 30–50, sometimes 100–150) and O^- centers toward the Fe^{3+} centers of luminescence (up to 30–60). The latter are distinguished by the maximum intensity of luminescence in practically ore-less granitic plutons, pegmatite bodies, inner zones of ore-bearing plutons, and peripheral zones of mineralized vein bodies. We add as well that the infrared centers of luminescence are above all typical (and moreover are distinguished by the greatest intensity) for amazonites from pegmatite and granite deposits with rare-metal mineralization.

Thus, the luminescence properties of amazonites in aggregate with the particularities of their color are fairly informative evaluation features of the character and degree of occurrence of amazonitization, the level of erosional exposure of amazonite-containing rocks, and the development in them of productive stages of diverse ore mineralization.

On the basis of all of the above, an explanation becomes possible for a range of previously incomprehensible facts (for example, the presence of oreless plutons of amazonitic granites and pegmatite vein bodies), which more than once have been supplied in the capacity of the primary evidence for the supposedly quite limited value of amazonite as a prospecting index (see Section 5.2). All of the results obtained by the author et al. [47] are persuasive, first of all that in oreless amazonite-containing granites and pegmatites, the properties and crystal chemistry of amazonite are totally different from those of amazonite occurring in association with ore mineralization. Precisely this aspect was neglected by those researchers who without any particular basis called amazonite an index for an entire complex of rare elements (K.K. Zhirov and S.M. Stishov), as well as by those skeptical of the advantages of the employment of amazonite in the capacity of a prospecting index [2]. Of course, in itself a find of amazonite in pegmatites or granites does not guarantee the discovery in the immediate vicinity of ore mineralization. This question can be answered only after conducting comprehensive complex research of amazonite; therefore, any discussion on the prospecting significance of this mineral will inevitably reach an impasse if it does not include analysis of the various data on the crystal chemistry and properties of amazonites from various genetic and formational types of granitoid rocks. In fairness, this situation has been informed by concrete data obtained in the research presented here for a range of granite and pegmatite deposits (see Table 6.1).

In conclusion, it follows to discuss one additional factor that complicates the use of amazonite in the capacity of a prospecting index. The fact is, in many granitic plutons or pegmatite veins, the occurrence or predominance has been established of only one (early or late) of the stages of the process of amazonitization. In these cases, it is all the more imperative to conduct detailed research on the amazonite itself. As an example, we will consider the pegmatites of the Kola Peninsula, where the oreless pegmatite veins most separated from the source pluton of alkaline granites contain pale-green amazonite ($\lambda = 550$–560 nm, $P = 10\%$, $Y = 40$–45%), while observed in the vein bodies with ore mineralization situated in the near exocontact of the granites are principally blue and greenish-blue varieties ($\lambda = 491$–500 nm, $P = 15$–20%, $Y = 25$–35%).

Yet another example of granitic plutons is as follows: in the practically oreless Maikul' pluton (Kazakhstan) predominate green amazonites ($\lambda = 520$–570 nm) of the relatively low-temperature late stage, whereas fairly widely developed in the Mesozoic pluton of eastern Siberia are rare-metal minerals and early blue and greenish-blue amazonites ($\lambda = 494$–505 nm) of the relatively high-temperature stage. A comparison of the crystal chemistry and properties of amazonites from these granite plutons shows that the composition of rare elements, the structure, the color, and the luminescence (see Chapter 4, Table 6.1) in each set of amazonites are highly specific.

Thus, all of the cited materials provide evidence of the important significance of the typomorphism of amazonite and of the promise of its use in the practice of prospecting-evaluation reconnaissance. The obtained research results on the crystal chemistry and properties of this mineral can be recommended in the capacity of support for the solution of a range of geological-genetic and practical problems arising with the prospecting and evaluation of rare-metal deposits associated with amazonite-containing rocks. Among them, attention should be directed above all on the potential of defining the following:

1. The formation types and subtypes of granitoid rocks
2. The level of erosion of plutons
3. The scale, degree, and type of ore bearing of the investigated geological objects

The resolution of these questions and problems should rely on the detailed analysis of the typomorphism of amazonite—its mineral associations, particularities of crystal chemistry, and properties. Moreover, a definite sequence should be observed when conducting various types of evaluations: from the simplest, undertaken in field conditions, to the progressively more complex, specialized methods of research used in laboratory conditions. The author et al. [47] recommend the following sequence: in the initial stage, fieldwork on finds of amazonite can examine its content, character of color, and mineral associations in various rocks to determine on first approximation the formation types (alaskite, subalkaline-leucogranite, alkaline-granite) and subtypes (greisen-albitite and amazonite-granite), level of erosion, intensity of ore formation, and petrogenic processes; subsequently, detailed study of the crystal chemistry and properties of amazonite can confirm conclusively all of the questions posed above, revealing the prospecting-evaluation value of this mineral and the scale and degree of ore bearing of concrete geological formations [16,46,47].

In summary, we must stress once again that it is important to recognize the possibilities that amazonite offers as a prospecting index for some types of rare-metals deposits (Ta, Nb), REE mineralizations, and some gemmological minerals (topaz, beryl). The contradictions in the current evaluations of these possibilities are the result of insufficient studies of the geological setting as well as of the crystal chemical and spectrometrical particularities of amazonite. As noted by Ostrooumov et al. [47], the differing geological, petrologic, and geochemical settings of amazonite crystals establish the distinctive features of its crystal chemical and spectrometrical particularities.

6.2 BRIEF HISTORY OF FINDS AND USE OF AMAZONITE AS A GEMMOLOGICAL STONE

The history of this semiprecious stone reaches back through the centuries. The first known employment of amazonite dates to the third to second centuries BC. According to the data of M. Blok, beads of amazonite have been found in Chaldea inside the royal tombs of Ur (ancient Sumer). This stone was also used in the adornment of chiefs of the ancient peoples of Central and South America. Finally, jewelry artifacts with amazonite have been discovered in the pyramids of ancient Egypt.

According to the data of A.E. Fersman [23,24], still in the epoch of Roman rule in Egypt, red porphyry and red granite, lamellar marbled onyx, and green feldspar were mined in vast quantities. In his notable "Essay about History stones" [25] he wrote the following regarding amazonite: "…In splendid fine amulets and beads we encounter it in Egypt, as though the name 'Wat' must partially be attributed to it. Whence acquired Egypt this stone, so close in tone to the turquoise minerals of the Sinai deposits or to the favored blue pastes. Where was its original deposit: in the upper Nile, in Ethiopia and contemporary Abyssinia, in the pegmatite veins of the granite desert of Libya, or in the eastern wilds of Kerman and Badakshan? The mineralogist cannot answer these questions, since he does not know pegmatite veins in these regions with amazonite, and pointing to the southern Urals or the veins of Mongolia can hardly provide the key to the splendid but rare stone of Egypt" [25, p. 242].

Already in the second half of the twentieth century, archaeologists together with geologists established deposits of green feldspar that had been developed by the Egyptians. Sites of the "quarries" of this stone were maintained on the territory of Egypt and neighboring Ethiopia. Unfortunately, it cannot yet be definitively stated from which deposits in ancient times amazonite was mined in America and Asia Minor. Likewise remains unknown the former name of this stone among the peoples of ancient civilizations; an answer to this question is found neither in the classical works of Theophrastus (third to second century BC), Pliny the Elder (first century BC, the first naturalists who composed tracts on stones) nor in the known writings of Al-Biruni (the most authoritative mineralogist of the Middle Ages). And although deposits of amazonite had been developed by the Egyptians and subsequently their Roman conquerors, this stone is not noted in the works of naturalists until the end of the eighteenth century, and it cannot be said whether it was known over the span of the many centuries separating us from the Roman Empire.

The name "amazonstone," in use even today, appeared at the end of the eighteenth century (J.-B. Romé de L'Isle, 1783), see in [25] and was applied initially to the several pitted and unreliably defined green pebbles imported from the Amazon basin by an unknown collector. From this collection, the stones of R.-J. Haüy were described as a light-green nephrite and green feldspar.

In Russia, amazonite was first discovered in the bedrock deposit in the pegmatite veins in the southern Urals in 1783. Interest toward the new stone was elicited by its range of beautiful colors; due to these, it recalled such rare and valuable stones as emerald, malachite, nephrite, and turquoise, Russian deposits of which at that time were not known. In the pegmatites of the southern Urals together with amazonite were found precious topazes, beryls, and phenakites. Occasionally here were observed also beautiful graphic intergrowths of amazonite with quartz. All of this together taken promoted the rapid growth of a wide popularity of amazonite in Russia and abroad. It augmented and adorned many private and museum collections of Europe.

However paradoxically, it was precisely the discovery of green feldspar in the Urals that contributed to the strengthening of the "amazonstone" appellation [9]. The beauty and uniqueness of the stone (at that time the only primary deposit) was striking not only to its discoverers. Promptly the instruction was issued to mine it for the Ekateringof lapidary factory and to manufacture vases from the highest quality stones. From Il'menskie amazonstone (more precisely from the rocks in which it was found in intergrowth with quartz-graphic granite), the master of the Petergof lapidary factory created four vases, which in 1794 adorned the collection of stone artifacts of the Hermitage museum. Later that stone began to be used in jewelry works, adorning pendants, brooches, cufflinks, buttons, and other decorative-artistic objects: cups, small vases, and caskets, as well as souvenirs and amulets. Additionally, the craftsmen of the Petergof lapidary factory via the "Russian mosaic" method created gala table tops for the palaces. Only after several decades did St Petersburg begin to receive from the Urals such renowned green stones as malachite and emerald, eclipsing before long the glory and beauty of amazonstone. In 1847, A. Breithaupt proposed the successful abridgement of the name of this mineral to "amazonite," which took hold in mineralogy.

The second half of the nineteenth century saw the discovery of new deposits of amazonite in the Americas, Madagascar, and elsewhere. Already in the twentieth century broad renown was acquired by amazonite thanks to the works of A.E. Fersman. "I never saw a more splendid image, despite having earlier managed to see many deposits of colorful stones … Nowhere have I been gripped by such a sense of admiration before the riches and beauty of nature as in the amazonite mines," thus he described his impressions from a visit to the deposits in the Il'menskie Mountains in the southern Urals [25, p. 188].

The increase in the number of deposits of this mineral gradually devalued amazonite, and consequently, it was relegated from the category of jewelry (used by royalty) to the category of essentially common ornamental stones. In contemporary classification [7], amazonite belongs to the group of jewelry-ornamental stones of the II order. Now, it is widely applied for the production of various ornaments and goods: cufflinks, pendants, brooches, insets for earrings, beads, rings, caskets, ashtrays, inkstands, lamps, table tops, etc. Additionally, fairly unique articles are known in which amazonite is used as a semiprecious stone, among them a panel map of France, which was displayed at the 1900 World's Fair in Paris, a stone map of the Soviet Union created at the end of the 1930s, and a map of the Il'menskie state reserve produced in 1978.

At present in Russia, the principal source of jewelry-ornamental amazonite is the Ploskaia Mountain deposit situated on the Kola Peninsula, unique in its occurrences of amazonites various in color (tone, saturation) and pattern (see some samples in Photos 7, 21, and 22). In the Urals, this mineral at the present is not mined, insofar as some veins have already been worked out, and others are situated on the territory of the Il'menskie state reserve where mining work is forbidden.

Thus, until recent times, from amazonite was obtained essentially various inexpensive stone artifacts. It evidently follows to number amazonitic granite among the semiprecious stones. Its deposits were first discovered in Russia and Kazakhstan by Russian geologists in the late 1930–1940s, but industrial use began only in the second half of the twentieth century [47]. Amazonitic granites serve as a beautiful facing stone and a construction material. The primary deposits at the present are Orlov and Etyka in Transbaikal, as well as Maikul' and Turangin in Kazakhstan. Granite of these deposits has been widely applied as a decorative and ornamental material in the construction of a range of buildings in certain cities of Russia and Kazakhstan. Occasionally for these purposes is used as well amazonitic pegmatite, from which have been created wall mosaics and interiors in a range of buildings in Moscow and other cities.

It follows to note that amazonite and amazonite-containing rocks can be considered also in the capacity of a collection material. For an illustration of the variety of natural amazonite at the present have been prepared small collections (10–15 specimens) from typical pegmatite and granite deposits (Kola Peninsula, Urals, Transbaikal and elsewhere).

References

[1] A. Vochmentzev, M. Ostrooumov, A. Platonov, Amazonite, Nedra, Moscow, 1989. 191 p. (in Russian).

[2] L. Feldman, A. Bugaets, V. Matias, Amazonitization of Granites in Relation to Ore Formation, Nedra, Moscow, 1967. 87 p. (in Russian).

[3] A. Bajtin, Rock-Forming Minerals: Optical Spectra, Crystalchemistry, Color, Typomorphism, Kazan university, 1985. 155 p. (in Russian).

[4] M. Beskin, V. Larin, Y. Marin, Granitoides Formations with Rare Metal Deposits, Nedra, Moscow, 1979. 247 p. (isn Russian).

[5] V. Vodopianova, A. Mirgorodski, A. Lazarev, Raman spectra of potassium feldspars, Inorg. Mater. 19 (11) (1983) 1891–1898 (in Russian).

[6] A. Zavaritsky, About amazonite, Proc. USSR Mineral. Soc. 44 (1) (1943) 29–38 (in Russian).

[7] E. Kievlenko, N. Senkevich, Geology of Gemstone Deposits, Nedra, Moscow, 1982. 324 p. (in Russian).

[8] M. Ostrooumov, Spectrometry investigation of the colour of amazonite, in: Modern Investigations in the Geology, Science, Leningrad, 1973, pp. 32–40 (in Russian).

[9] M. Ostrooumov, Amazonstone, Nature 3 (1982) 45–53 (in Russian).

[10] M. Ostrooumov, Colorimetry of amazonite, Proc. USSR Mineral. Soc. 116 (1) (1987) 437–448 (in Russian).

[11] M. Ostrooumov, I. Musina, Evolution of amazonite perthite color in Ploskaya pegmatite (Kola peninsula), Proc. USSR Mineral. Soc. 104 (6) (1975) 738–741 (in Russian).

[12] M. Ostrooumov, A. Platonov, A. Taraschan, On the nature of colour of amazonite, in: Physics and Typomorphism of Minerals, Science, Leningrad, 1976, pp. 52–61 (in Russian).

[13] M. Ostrooumov, A. Vochmentzev, Y. Marin, Typochemical characteristics of amazonite, Proc. USSR Mineral. Soc. 110 (4) (1981) 437–448 (in Russian).

[14] M. Ostrooumov, A. Vochmentzev, Y. Marin, Structural state of amazonite, Proc. USSR Mineral. Soc. 111 (6) (1982) 719–733 (in Russian).

[15] M. Ostrooumov, A. Vochmentzev, Colorimetry of minerals, Nature 6 (1987) 43–53 (in Russian).

[16] M. Ostrooumov, A. Platonov, Typomorphic significations of the amazonite, Mineral. J. 10 (3) (1988) 3–11 (in Russian).

[17] M. Ostrooumov, Method of determination of the lattice ordering degree in the alkaline feldspars by infrared reflection spectrometry, Proc. USSR Mineral. Soc. 5 (1991) 94–99 (in Russian).

[18] V. Popov, V. Popova, V. Polyakov, Pegmatites of Ilmen Mountains, Ural Science Center of Russian Academy, 1982. 127 p. (in Russian).

[19] A. Platonov, Color Nature of Minerals, Naukova Dumka, Kiev, 1976. 264 p. (in Russian).

[20] A. Platonov, A. Tarachan, M. Taran, About color centers in amazonites, Mineral. J. 6 (4) (1984) 3–16 (in Russian).

[21] V. Popov, M. Ostrooumov, Morphological observations of the amazonite, in: Onthogenuos Observations in the Geology, Science, Leningrad, 1984, pp. 82–91 (in Russian).

[22] S. Projor, M. Ostrooumov, Colorimetry of the nephrite, Mineral. J. 14 (3) (1992) 78–85 (in Russian).

[23] A. Fersman, Pegmatites, vol. 6, Russian Academy of Science, 1960 (in Russian).

[24] A. Fersman, Travels for Stones, Gosizdat, 1956 (in Russian).

[25] A. Fersman, Essay about History Stone, vol. 1, Russian Academy Science, 1954 (in Russian).

[26] T.S. Ercit, REE-enriched granitic pegmatites, in: R.L. Linnen, M.L. Samson (Eds.), Rare-element Geochemistry and Mineral Deposits, GAC Short Course Notes, vol. 17, Geological Association of Canada, 2005, pp. 175–199.

[27] E. Foord, R. Martin, Amazonite from the Pikes Peak Batholite, Mineral. Rec. 10 (1979) 373–374.

[28] A. Hofmeister, G. Rossman, A spectroscopic study of irradiation coloring of amazonite: structurally hidrous Pb-bearing feldspars, Am. Mineral. 70 (7–8) (1985) 794–804.

[29] A. Julg, A theoretical study of the absorption spectra of Pb^+ and Pb^{3+} in site K^+ of microcline: application to the color of amazonite, Phys. Chem. Miner. 23 (3) (1998) 229–233.

[30] A. Marfunin, Advanced Mineralogy, Springer-Verlag, Berlin, 1995.

[31] R.F. Martin, K. De Vito, F. Pezzotta, Why is amazonitic K-feldspar an earmark of NYF-type granitic pegmatites? Clues from hybrid pegmatites in Madagascar, Am. Mineral. 93 (2008) 263–269.

[32] I. Oftedal, Heating experiments of amazonite, Mineral. Mag. 31 (1957) 236.

[33] M. Ostrooumov, E. Fritsch, B. Lasnier, Infrared reflection spectrometry of minerals and gemstones, in: M. Pecchio, H. Kahn (Eds.), Developments in Science and Technology. Applied Mineralogy, vol. 2, 2004, pp. 595–598.

[34] M. Ostrooumov, L' amazonite, Revue de Gemmologie 108 (1992) 8–12.

[35] M. Ostrooumov, Mineralogía Avanzada en México: conceptos, resultados, investigaciones futuras, Boletín de la Sociedad Mexicana de Mineralogía 14 (2001) 7–16.

[36] M. Ostrooumov, Espectrometría infrarroja de reflexión en Mineralogía Avanzada, Gemología y Arqueometría: Monografías del Instituto de Geofísica, Universidad Nacional Autónoma de México, 2007. 87 pp.

[37] M. Ostrooumov, First discovery of amazonite in Mexico, Gems Gemol. 43 (2007) 163–164.

[38] M. Ostrooumov, Mineralogía Analítica Avanzada: Universidad Michoacana de San Nicolás de Hidalgo-Sociedad Mexicana de Mineralogía, 2009. ISBN: 978-607-424-095-5. 275 pp.

[39] M. Ostrooumov, Avances recientes de la espectroscopía Raman en Ciencias de la Tierra, Serie Monografías, vol. 18, Universidad Nacional Autónoma de México, 2012. ISBN: 978-607-02-3924-3. vol. 2, 250 p.

[40] M. Ostrooumov, Algunas consideraciones mineralógicas y geoquímicas sobre el hallazgo de amazonita en el estado de Chihuahua, México, Revista Mexicana de Ciencias Geológicas 29 (1) (2012) 221–232.

[41] M. Ostrooumov, B. Lasnier, S. Lefrant, Spectrometrie infrarouge de reflexion des mineraux et materiaux gemmes, Analusis 23 (1) (1995) 39–44.

[42] M. Ostrooumov, E. Fritsch, S. Lefrant, B. Lasnier, Spectres Raman des opales: aspect diagnostique et aide a la classification, Eur. J. Mineral. 11 (5) (1999) 899–908.

[43] M. Ostrooumov, E. Fritsch, A. Victoria Morales, Algunas consideraciones sobre la naturaleza del color de topacios volcanicos (Mexico), Boletin de la Sociedad Española de Mineralogía 24 (2001) 33–41.

[44] M. Ostrooumov, E. Faulques, E. Lounejeva, Raman spectrometry of natural silica in Chicxulub impactite (Mexico), Comptes Rendus de l'Academie des Sci. 334 (2002) 21–26.

[45] M. Ostrooumov, E. Fritsch, E. Faulques, Etude spectrometrique de la lazurite du Pamir, Can. Mineral. 40 (3) (2002) 885–893.

[46] M. Ostrooumov, A. Banergee, Amazonite from Pre-cambrian pegmatites (Kola Peninsula, Russia): crystal chemical and spectrometric study, Schweiz. Mineral. Petrogr. Mitt. 85 (1) (2005) 89–102.

[47] M. Ostrooumov, A. Platonov, V. Popov, Amazonstone: Mineralogy, Crystal Chemistry, Typo-morphism, Polytechnics, St. Petersburg, 2008. ISBN: 978-5-7325-0675-4. 255 p. (in Russian).

[48] I. Petrov, R. Mineeva, L. Bershov, EPR of [Pn–Pb]$^{3+}$ mixed valence pairs in amazonite-type microcline, Am. Mineral. 78 (1993) 5–6, 500–510.

[49] B. Speit, J. Lehman, Radiation defects in Feldspars, Phys. Chem. Miner. 8 (1982) 77–82.

[50] A. Szuzkiewicz, T. Körber, "Amazonit" oder "Gruner microcline"? Zur Ursache der Grünfär-bung von Kalifeldspäten aus dem Striegauer Granit, Lapis 35 (7–8) (2010) 75–77, 86.

[51] A. Voloshin, Y. Pakhomovsky, Mineralogical Evolution in Pegmatites of Kola Peninisula, Nauka, Leningrad, 1986. 168 pp. (in Russian).

Index

Note: Page numtbers followed by "f" indicate figures, "t" indicate tables.